Electronic Components

A Complete Reference for Project Builders

Electronic Components
A Complete Reference for Project Builders

Delton T. Horn

TAB BOOKS
Blue Ridge Summit, PA

FIRST EDITION
FIRST PRINTING

© 1992 by **TAB Books**.
TAB Books is a division of McGraw-Hill, Inc.

Library of Congress Cataloging-in-Publication Data

Horn, Delton T.
    Electronic components : a complete reference for project builders
/ by Delton T. Horn.
        p.   cm.
    Includes index.
    ISBN 0-8306-3335-9 (h)   ISBN 0-8306-3333-2 (p)

    1. Electronics—Amateurs' manuals.   2. Electronic circuits-
-Amateurs' manuals.   I. Title.
TK9965.H65   1991
621.381—dc20   91-11930
                                                        CIP

TAB Books offers software for sale. For information and a catalog, please contact TAB Software Department, Blue Ridge Summit, PA 17294-0850.

Acquisitions Editor: Roland S. Phelps
Book Editor: Laura J. Bader
Production: Katherine G. Brown
Book Design: Jaclyn J. Boone
Managing Editor: Sandra L. Johnson
Paperbound cover photo: Brent Blair Photography, Harrisburg, PA
Paperbound cover design: Lance Bush, Hagerstown, MD   EL2

Contents

❖ **Part 2** ❖
Active components

❖ **Part 3** ❖
Miscellaneous components

Preface

THIS BOOK IS INTENDED AS A SINGLE-SOURCE HANDBOOK FOR the electronics experimenter and technician, compiling information on a wide variety of common electronic components. Most of this information is available from other sources, of course, but here it is gathered in one convenient reference.

You can read this book as an introductory or refresher text, or you can simply look up the type of electronic component you are interested in the table of contents and quickly have the relevant information right at your fingertips.

The functioning of basic electronic components is covered, along with the differences between the various subtypes of each major class of component. The important factors to consider when making substitutions in existing circuits are also discussed.

Just a glance at the table of contents will give you an idea of the range of topics covered in this volume. The book is divided into three major sections. Part 1 deals with passive, nonamplifying components, beginning with the simple wire and solder that is part of every electronic circuit (chapter 1). This is followed by an in-depth examination of the three major types of passive components and related devices—resistors (chapter 2), capacitors (chapter 3), and inductors (chapter 4). Specific types within each of these broad categories are covered. For example, there are many different types of capacitors, yet few electronic hobbyists know which type to use in what kind of circuit or which types are interchangeable and which aren't.

Part 2 covers active devices, which are capable of amplification. The focus here is entirely on semiconductor components. This includes diodes (chapter 5) and transistors (chapter 6) of many different types, including zener diodes, four-layer diodes, LEDs, laser diodes, UJTs, FETs, and SCRs, among others. We will also look at integrated circuits (ICs), although practical limitations restrict the extensiveness of this section. Chapter 7 deals with linear ICs, such as amplifiers, op amps, timers, and more. Digital ICs are covered in chapter 8, ranging from simple gates to common digital devices, such as flip-flops, counters, and shift registers.

Part 3 includes some miscellaneous devices that don't really fit into the other categories. Chapter 9 discusses transducers, which convert some external condition into an electrical signal or vice versa. Finally, chapter 10 discusses switches.

Of course, it would be impossible to discuss every electronic component available. There are literally millions, and more are being developed every week. This is especially true in the case of ICs. This book discusses major component types, dealing with specific components only as examples of more general principles.

Armed with this book and a manufacturer's spec sheet, the electronics experimenter or technician can get the most out of almost any electronic component.

❖ Part 1
Passive components

Wire and solder

SOME READERS MIGHT FIND IT A LITTLE ODD THAT A BOOK ON electronic components would begin with a chapter on wire and solder. Surely those things aren't actual components, are they? Yes, in a way they are. They are an essential part of every electronic circuit. If they are not properly selected and used, the circuit might not function as desired. Wire and solder might be rather mundane, but they are not unimportant topics.

The basics of wire

The main function of wire in an electric or electronic circuit is to conduct electricity from one place to another. Component leads are a form of wire. Often in practical circuits, some additional wire will be required. In some cases, wire might be used for physical reinforcement. Special types of wires are used to make lamp filaments, heating elements, wire-wound resistors (see chapter 2), and coils (see chapter 4).

Basically, a wire is a long, narrow cylindrical piece of conductive metal. Common metals used in wires include copper, silver, steel, iron, aluminum, and even gold. Most standard wire is made from copper because it is the most efficient conductor for a reasonable cost. In some cases silver or gold wires might do a better job than a copper wire, but such metals are far more expensive.

Insulation

Sometimes a wire will be bare, especially if it is used as a short jumper. In many practical cases, however, bare wires are highly undesirable. If two (or more) bare wires (including component leads) touch, a short circuit can occur. That is, the electrical current won't go where it's supposed to go, but instead it goes someplace where it is unwanted.

To prevent such shorts, most wires (especially if more than an inch or two long) are insulated. This means the conductive wire is enclosed in a nonconductive (insulator) shell. Rubber, certain types of fabric, nylon, and various plastics are often used as insulation on wires.

Remember, no practical insulator is perfect. If very large currents are flowing through a wire, some of the current can leak through the insulation. This is why it is dangerous to touch any ac power lines, even if they are insulated.

An insulated wire is first cut to length, then a small amount of insulation is stripped off the ends of the wire. Usually about 0.5 in. to 0.75 in. of insulation will be removed from each end. The bare end of the wire is used to make the appropriate circuit connection by soldering it in place. The insulation cannot be soldered. At best, it will just melt and make a mess. However, some of the melted insulation could easily get in the way of the desired joint connection, preventing good electrical contact. Some fabric insulations may be slightly flammable, while some plastics may cause dangerous fumes if they are melted.

There is one exception to this rule. Some wires used for wire wrapping (a specialized type of circuit construction) use a special vaporizing insulation. When heated, this insulation vaporizes into harmless fumes. A solder joint can be made in the middle of a length of wire. However, when in doubt, assume that the insulation must be removed before any soldering can be done.

Insulation can be stripped off of a wire with a pocket knife, but it is important to avoid nicking the wire. A nicked wire is mechanically weak and may break at the weakened point. Special wire-stripping tools are available. One of the most popular and inexpensive types of wire stripper is a very simple pliers-like device with notched shear-type blades. The wire is placed in the notch and the jaws of the tool are held closed as the wire is pulled through. The notch blades cut into the insulation and yank it off as the wire is pulled through the narrow notch. The notch can be manually set for various widths to accommodate different sizes of wire. Often it is necessary to use a pair of long-nose pliers to hold the wire securely with one hand, while using the wire stripper with the other hand.

A variation on this device is a combination tool used for both wire stripping and for crimping solderless terminals. Not unexpectedly, this more versatile tool tends to be a bit more expensive. There are some household uses for solderless connectors, so it might not be a bad investment.

The best and most efficient type of wire stripper is a rather nightmarish-looking pair of specialized pliers with a viselike arrangement in one set of jaws to hold the body of the wire and a set of knife edges in the other jaw with notches in them to fit different sizes of wire. When the handles of this tool are squeezed together, the wire is gripped snugly by the vise jaws and pulled away from the knife jaws, which cut into and neatly strip away the insulation. This type of semiautomatic tool is much more expensive than a simple manual wire stripper, but it is faster, more convenient, and does a better job. A semiautomatic wire stripper like this never nicks wires. It is also fascinating to watch it in action. The price range for such wire-stripping tools runs from about $7 to more than $25, depending on the size and sophistication of the design.

Some wires are insulated with a thin coating of enamel. Such wire is called enameled wire, for obvious reasons. This type of insulation is very inexpensive and doesn't add much to the diameter of the wire. Other types of insulation can be quite thick. Unfortunately, enamel insulation is much more difficult to remove than other forms of insulation. Wire strippers, like those described above, probably won't work very well, if at all. Usually the enamel must be scraped off the wire with a knife or razor blade.

Enameled wire is most often used in making coils (see chapter 4) and wire-wound resistors (see chapter 2) where many closely spaced turns of wire must be electrically insulated from each other. Other types of insulation might make the wire too thick for this purpose.

Wire sizes

Wire is available in a number of sizes. The size refers to the diameter of the conductor (without the insulation). The maximum current a given wire can safely carry is determined by the conductor material used and the diameter of the wire. Generally speaking, the thicker the wire (larger the diameter), the greater the maximum current flow the wire can handle. This makes sense—there's "more room" for electrons to flow through a thicker wire.

The diameter of a wire can be expressed in inches (usually fractions of an inch), millimeters, or some other linear unit of measurement. Because the wires used in most electronic circuits are so thin, a very useful linear unit of measurement is the mil. One mil is equal to 0.001 in., or about 0.0254 mm. The mil isn't in common use today, but you may encounter the term occasionally, especially in reference to wire sizes.

In most electronics work, however, the size of wire is defined by a standard gauge number. There are three wire gauge standards in common use. They are the American Wire Gauge, the British Wire Gauge, and the Birmingham Wire Gauge. In the United States, the American Wire Gauge is most commonly used. This standard is often abbreviated AWG.

AWG values range from 1 to 40. The larger the AWG number, the thinner the wire. Table 1-1 lists the diameters (in millimeters) for each of the 40 standard AWG sizes. The diameter refers to the actual conductor only. Any insulation layers are not included in the gauge size because the thickness of the insulation does not affect the wire's current-handling capability.

Assuming the conductor material is not changed, decreasing the AWG number indicates an increase in the current-handling capability. Thick wires have low AWG numbers, and thin wires have high AWG numbers. The odd-number AWG sizes don't seem to be in very widespread use.

For general purpose low-power electronics work, AWG 22 wire is often the best choice. It is light and flexible and doesn't eat up a lot of space, yet it is large enough to be sturdy and reliable. This size wire can also conduct a reasonable amount of current without problems.

For high-power applications, a heavier gauge wire may be required. Some electronic circuits may use AWG 20, AWG 18, AWG 16, or even AWG 14 wire.

Table 1–1   American Wire Gauge system sizes.

AWG number	Wire diameter (mm)	AWG number	Wire diameter (mm)
1	7.35	21	0.723
2	6.54	22	0.644
3	5.83	23	0.573
4	5.19	24	0.511
5	4.62	25	0.455
6	4.12	26	0.405
7	3.67	27	0.361
8	3.26	28	0.321
9	2.91	29	0.286
10	2.59	30	0.255
11	2.31	31	0.227
12	2.05	32	0.202
13	1.83	33	0.180
14	1.63	34	0.160
15	1.45	35	0.143
16	1.29	36	0.127
17	1.15	37	0.113
18	1.02	38	0.101
19	0.912	39	0.090
20	0.812	40	0.080

For very low-power applications, a thinner gauge wire may be used, particularly when the physical space of the circuit must be kept as compact as possible. Wire-wrapping wire is usually AWG 30.

In England and certain other countries the British Wire Gauge system is normally used instead of the AWG system. The abbreviation NBS SWG is commonly used to indicate the British Wire Gauge system.

Like the AWG system, the NBS SWG system employs 40 gauge sizes from 1 to 40. The conductor diameter decreases as the gauge number increases. Only the conductor diameter is measured. The thickness of the insulation (if any) is ignored in determining the gauge size.

The NBS SWG sizes are defined in terms of inches, rather than millimeters. This system is summarized in Table 1-2. The differences between the AWG and NBS SWG sizes are slight and can usually be ignored.

A third size system used in some parts of the world is the Birmingham Wire Gauge, abbreviated as BWG. Only 20 wire sizes (from 1 to 20) are identified in this system, which is summarized in Table 1-3.

Table 1 – 2 British Wire Gauge system sizes.

NBS SWG number	Wire diameter (in.)	NBS SWG number	Wire diameter (in.)
1	0.300	21	0.032
2	0.276	22	0.028
3	0.252	23	0.024
4	0.232	24	0.022
5	0.212	25	0.020
6	0.192	26	0.018
7	0.176	27	0.0164
8	0.160	28	0.0148
9	0.144	29	0.0136
10	0.128	30	0.0124
11	0.116	31	0.0116
12	0.104	32	0.0108
13	0.092	33	0.0100
14	0.080	34	0.0092
15	0.072	35	0.0084
16	0.064	36	0.0076
17	0.056	37	0.0068
18	0.048	38	0.0060
19	0.040	39	0.0052
20	0.036	40	0.0048

Table 1 – 3 Birmingham Wire Gauge system sizes.

BWG number	Wire diameter (mm)
1	7.62
2	7.21
3	6.58
4	6.05
5	5.59
6	5.16
7	4.57
8	4.19
9	3.76
10	3.40
11	3.05
12	2.77
13	2.41
14	2.11
15	1.83
16	1.65
17	1.47
18	1.25
19	1.07
20	0.889

The BWG numbers are not the same as in the AWG or NBS SWG systems, but the actual wire sizes are similar. As usual, the gauge size defines only the diameter of the actual conductor in the wire. The thickness of the insulation (if any) is ignored.

Color-coding wires

Many professional technicians and serious electronics hobbyists routinely color code the wires in their circuits and projects. Hookup wire is available with insulation in a wide variety of colors. Of course the color of the insulating jacket on your project's wires will have no effect on the performance of the circuit.

But color-coded wires will result in a neater layout that is much easier to troubleshoot, especially if large bundles of wires are involved. The color of the wire's insulation will give a clue to what it is connected to in the circuit.

There is no official standard color-coding system for wires, but the system outlined in Table 1-4 is widely used.

Table 1 – 4   Standard color – coding for wires.

Color	Use
Black	All grounds
Brown	Heaters or filaments off ground
Red	Positive dc supply voltage
Orange	Screen grids of tubes; base 2 of transistors; gate 2 of FETs
Yellow	Transistor emitters; FET sources; tube cathodes
Green	Transistor bases; FET gates; tube control grids; ac commons
Blue	Transistor collectors; FET drains; tube plates
Violet	Negative dc supply voltage
Gray	ac Supply leads
White	Bias and AGC lines

Solid and stranded wire

There are two basic types of wire: solid wire and stranded wire. In a solid wire, the conductor is a continuous length of metal. There is only one conductor. A stranded wire, however, is made up of many very thin wires wound around each other to form the diameter of the wire.

Solid wire is easy to wind around a joint and is convenient to solder. With stranded wire, some individual wires may tend to get away from you, causing an undesired short circuit if you're not careful. Stranded wire also tends to suck up molten solder. In fact, a piece of heavy-gauge stranded wire is sometimes used as desoldering braid. To make a good solder joint with stranded wire, you should tightly twist the strands at the end of the wire and tin the tip. This means apply a little solder to the end of the wire before mechanically affixing it to the joint to be soldered.

Solid wire is inclined to break easily if it is nicked while the insulation is being removed. Solid wire is also less flexible than stranded wire of a similar gauge. If you bend a length of solid wire back and forth too sharply or too often, it will eventually break. This is less likely to be a problem with stranded wire. The individual strands of conductor tend to reinforce each other, and if one or two strands do break, most of the conductor will remain intact. Even the broken strands won't be completely disconnected because they are electrically shorted to the rest of the strands. Current will continue to flow through the stranded wire unless all (or most) of the strands are severed.

Multiconductor wire

So far we have dealt only with single-conductor wires. That is, the wire consists of one conductor wrapped in an insulation jacket, as illustrated in Fig. 1-1. Only a single voltage can be carried by such a wire at any given instant.

Some wires, however, contain two or more individual conductors that are internally insulated from each other. There are two basic ways to do this.

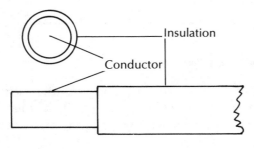

Insulation

Conductor

Fig. 1-1 *Standard hookup wire has just one conductor.*

Some multiconductor wires are really just a group of several individually insulated single-conductor wires wrapped in a common jacket, as shown in Fig. 1-2. Usually, the insulation on the individual wires is of different colors to permit convenient color coding of functions.

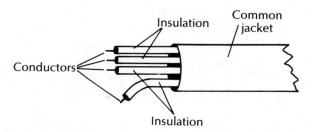

Fig. 1-2 *Some multiconductor wires consist of several regular hookup wires in a common jacket.*

Other multiconductor wires use separate wires with joined insulation, as illustrated in Fig. 1-3. Usually, this type of arrangement is used for two conductors. It is commonly called zip cord, presumedly for the sound made when the two conductors are

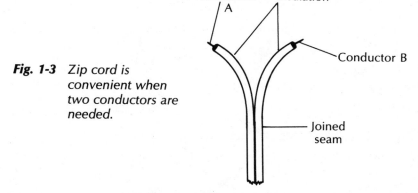

Fig. 1-3 *Zip cord is convenient when two conductors are needed.*

pulled apart for stripping and connecting to individual joints. Zip cord is used in electrical extension cord and in speaker wire, among other applications.

While a zip cord has just two conductors, the idea can be extended to over two dozen individual conductors. When many conductors are joined in this manner, the combination is known as a ribbon conductor or a ribbon cable.

Cables

Any grouping of electrical conductors bound together is known as a cable. The multiconductor wires described in the preceding section of this chapter are all cables of a sort.

Generally, in common practice, the term cable is reserved for relatively heavy cords, usually covered by a tough protective insulator. Rubber and polyethylene are widely used for this purpose. This thick jacket protects the conductors from external environmental conditions, especially moisture, and this helps prevent corrosion of the conductors within the cable. Unlike the relatively thin insulation normally used on ordinary hookup wires, most cable jackets are thick, durable, and designed for either indoor or outdoor use. Each conductor within the cable is individually insulated, of course.

The most familiar and common type of true cable is the coaxial cable used in many radios. This is often referred to as coax. A coaxial cable usually contains two conductors. The construction of a typical coaxial cable is illustrated in Fig. 1-4. The center conductor is a regular wire, usually of a fairly heavy gauge. This center conductor is surrounded by an insulating

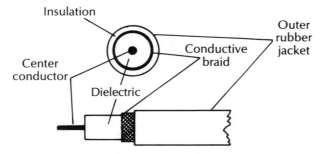

Fig. 1-4 *Coaxial cable is frequently used in radio applications.*

material known as the dielectric. Surrounding the dielectric layer is a web of thin woven wires that acts as the cable's second conductor. In effect, these braided wires are similar to a stranded wire. This outer conductor is usually employed as the ground connection. This surrounding outer conductor effectively shields the inner, center (signal) conductor from most external interference. A coaxial cable also offers relatively low signal loss because of this built-in shielding. Some coaxial cables have two or more center conductors that are electrically insulated from each other.

A coaxial cable bears some resemblance to a capacitor (see chapter 3). A capacitor consists of two conductors separated by an insulating material called a dielectric. That is exactly the situation we have with a coaxial cable.

Because of its capacitive nature, a coaxial cable has a characteristic impedance. Impedance is ac resistance, which varies with the signal frequency. Impedance is discussed in some detail in chapter 4. Coaxial cables can be designed for several different characteristic impedances. Two common types of coaxial cable have impedances of 75 Ω (used for TV antennas) and 50 Ω (used for CB radio aerials).

Fiber optics

There is one special type of cable that is used in electronic circuits, but it is not an electrical conductor. It is a fiber-optic cable, or as it is sometimes called, an optical cable.

A fiber-optic cable is made of spun glass or some transparent plastic. Instead of conducting electricity, it conducts encoded light signals, even along a path that twists and turns around multiple obstacles.

A fiber-optic system is almost totally immune to interference and noise, and a great many individual signals can be multiplexed and carried along a single optical fiber. Also, because there is no electrical current involved, there is never any shock or fire hazard as the signal is conducted from one place to another.

At the input end of a fiber-optic cable, a small laser diode (see chapter 5) is focused on the end of the relevant fiber. The laser light is modulated by the signal to be transmitted. In other words, the electrical signal is converted into a light signal. At the receiving end of the fiber-optic cable is a photosensor (see chapter 9) that converts the modulated light signal back into electrical form.

Types of solder

Solder is used to create a good, solid mechanical and electrical connection at each joint between two or more component leads or wires. Without solder the connections would be unreliable and circuit performance would be erratic.

Basically, solder is a metal alloy with a very low melting point. It is melted over a joint with a soldering iron that has a tip at a high temperature. The molten solder is permitted to flow over the joint and then cool. When it cools, it solidifies, fusing the leads at the joint together. If done properly, soldering forms a strong mechanical and electrical connection between the leads.

Most types of solder are an alloy of lead and tin. The relative ratio of these two metals determines the specific melting temperature of the solder. Occasionally, other metals, such as silver, might be used.

In most practical electronics work, 60-40 solder is used. This type of solder is comprised of 60% tin and 40% lead, and has a fairly low melting temperature. It can be melted without damage to heat-sensitive components in the circuit. Semiconductor components such as diodes, transistors, and ICs (see chapters 5 through 8) are particularly heat sensitive and can be damaged or destroyed if the hot tip of the soldering iron is applied too long. Other common types of solder are summarized in Table 1-5.

Table 1–5 Typical types of solder.

Metal used	Core	Melting point °F	°C	Typical applications
Tin-lead 50–50	Rosin	430	220	Electronics (mostly tube circuits)
Tin-lead 60–40	Rosin	370	190	Electronics (general purpose)
Tin-lead 63–37	Rosin	360	180	Electronics (low heat)
Tin-lead 50–50	Acid	430	220	Metal bonding (not for electronics use)
Silver	—	600	320	High heat, high current

Most types of solder are in a wirelike form, wound on a spool. Usually, there will be a core of some sort. This core material flows over the joint when the solder is melted, cleaning it and preparing it for a good electrical connection. For electronics solder, a rosin core is used. Some solder is solid, with no core. To use such solder, the rosin must be applied separately. The rosin is usually in the form of a paste.

Some solders designed for metal-working applications use

an acid core. Never use acid-core solder in any electronic circuit. The acid is highly corrosive and will eat through many of the components and, possibly, the circuit board. Acid-core solder is used only for metal bonding applications.

While most solder is sold on a spool in wirelike form, some electronics technicians and hobbyists prefer to use solder paste. Solder in paste form is messy and somewhat awkward to apply, but it has a lower melting point and can be used in tight spaces where wire solder would be inconvenient. A separate rosin paste must usually be employed along with solder paste.

One final type of solder is tape solder. This type of solder comes in small strips, often with a slightly adhesive surface. The solder tape is wound around the joint, then heat is applied. Tape solder melts at a very low temperature. In fact, a tape solder joint can be made with an ordinary match or cigarette lighter. This is the sole advantage of tape solder. At best, the solder joints made with tape solder are only fair. It is difficult to get the tape solder to melt and flow just right. It is also a little tricky to make a tape solder joint without burning your fingers.

Generally, tape solder should be used only for emergency repairs, not for general soldering. You certainly wouldn't want to construct an entire project with tape solder. It would be expensive (tape solder generally costs more than other forms of solder), and it would be extremely inconvenient and time-consuming. When making multiple solder joints, you can do a better job with a soldering iron. Use tape solder only in a pinch when nothing better is available. Tape solder does come in handy for unexpected emergency repairs, so it is a good idea to keep a package on hand, just in case.

❖ 2
Resistors

THE MOST BASIC AND COMMON OF ALL ELECTRONIC COMPONENTS is the resistor. A resistor is simply a device that offers resistance to the flow of electric current. Some of the electrical energy is "used up" as it passes through the resistor. Actually, this energy is dissipated by the body of the resistor in the form of heat.

In a sense, all electronic components, including the interconnecting wires, are resistors. No conductor is perfect. Some of the electrical energy will always be consumed and dissipated as heat. Different substances have differing amounts of resistance. A conductor has a very low resistance, and an insulator has a very high resistance. A semiconductor, not surprisingly, has a resistance somewhere between that of a conductor and that of an insulator.

Electrical resistance is somewhat analogous to mechanical friction. Friction resists, or slows down, mechanical motion. Similarly, resistance resists, or slows down, electrical motion (current flow). In some cases, mechanical friction is undesirable and wastes energy. In other cases, mechanical friction is highly desirable or even essential. Would you really want to drive a car if there was no friction to make the brakes work or keep the wheels on the road? Electrical resistance is sometimes a wasteful nuisance, but sometimes it is an essential element in the desired functioning of the circuit.

While all electronic components offer some resistance to the flow of electric current, the term resistor is normally reserved for a component that is designed to present a specific and predictable amount of resistance to the applied electrical current. A resistor's primary function is to offer resistance to the flow of electric current.

This may sound like a very silly, useless, and wasteful component. Why would we ever want to deliberately waste some of the electrical energy in a circuit, just to dissipate it as waste heat? Actually, resistors have countless applications in practical electronic circuits. Almost every electronic circuit includes at least one resistor. The resistor is the most commonly used of all electronic components. The standard symbol for a resistor is a zigzag line, as shown in Fig. 2-1.

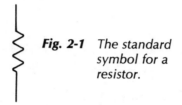

Fig. 2-1 *The standard symbol for a resistor.*

The ohm

The basic unit of measurement of resistance is the ohm. If 1 V is applied across a 1-Ω resistor, a current of 1 A will flow through the resistor. The Greek letter omega (Ω) is commonly used to represent ohms. For example, 15 ohms might be written as 15 Ω.

For many practical purposes, an ohm is too small a value, so larger units are often used. A kilohm (kΩ) is equal to 1,000 Ω. That is, a 2.2-kΩ resistor has a resistance of 2,200 Ω. A still larger resistance value is the megohm, which is often abbreviated as MΩ. One megohm is equal to 1,000,000 Ω (or 1,000 kΩ). For instance, a 3.9 MΩ resistor has a resistance of 3,900,000 Ω (or 3,900 kΩ).

In performing any mathematical equations involving resistance, make sure that all of the resistance values are used in the same unit form. That is, all of the resistances in the equation should be in ohms, or all in kilohms, or all in megohms. Do not mix the value types in a single equation.

Most standard formulae assume that resistance values will be in ohms, but it often won't matter which type of value you use, as long as you are consistent throughout the equation.

Occasionally, you may come across a reference to conductivity, especially in discussing the efficiency of a conductor. Conductivity is the reciprocal of resistance and is measured in mhos (ohms spelled backwards):

$$\text{mhos} = 1/\text{ohms}$$

or

$$\text{ohms} = 1/\text{mhos}$$

Conductivity is used to express awkwardly small resistances. For example, if you're dealing with an extremely good conductor with a resistance of 0.00004 Ω, it's a lot easier to say the conductor has a conductivity of 25,000 mhos.

Ohm's law

The ohm was named in honor of a German scientist, Georg Simon Ohm (1787 – 1854). Ohm was a very influential figure in the early development of the field of electronics. His most important contribution is known as Ohm's law. This is the single most important formula in all of electronics.

Ohm's law defines the interlinking relationship between voltage, current, and resistance at any point in any electrical circuit. This simple but powerful formula is

$$E = IR$$

where
E = voltage, in volts;
I = current, in amperes; and
R = resistance, in ohms.

For example, if we have 2.5 A of current passing through a 500-Ω resistor, the voltage is equal to

$$E = 2.5 \times 500$$
$$= 1,250 \text{ V}$$

Most practical electronic circuits use much smaller currents than this. Usually the current in an electronic circuit is measured in milliamps (thousandths of an ampere) or microamps (millionths of an ampere). The current value must be converted to amperes in order for the Ohm's law equation to work properly.

As an example, let's say we're still using a 500-Ω resistor, but now the current is just 12 mA (milliamps). Since 1 mA is equal to

0.001 A, the current must be rewritten as 0.012 A to be used in the Ohm's law formula. In this case, the voltage works out to

$$E = 0.012 \times 500$$
$$= 6.0 \text{ V}$$

Often, we will know the voltage and the resistance and we need to solve for the unknown current. It is a simple matter to rearrange the basic Ohm's law formula:

$$E = IR$$
$$I = \frac{E}{R}$$

For instance, let's say we apply 4.5 V across a 2200-Ω (2.2-kΩ) resistor. The current flowing through the resistor will have a value of

$$I = \frac{4.5}{2200}$$
$$= 0.002 \text{ A}$$
$$= 2 \text{ mA}$$

Notice that the result was rounded off in this example.

A third possibility is when we know the voltage and the current and we need to determine the resistance. This will most often occur in circuit design, when we have to select the appropriate resistor to give the desired results. Once again, it's just a matter of rearranging the basic Ohm's law formula:

$$E = IR$$
$$R = \frac{E}{I}$$

As an example, let's say we have 9 V and need a current flow of 32 mA (0.032 A). The appropriate resistor value works out to approximately

$$R = \frac{9}{0.032}$$
$$= 281.25 \text{ } \Omega$$

Unless the application is very critical, this can be rounded off to 270 Ω, which is the nearest standard resistor value. We will discuss standard resistor values later in this chapter.

The Ohm's law formula will work if R (resistance) is in kilohms and I (current) is in milliamps. But be careful not to combine kilohms and amperes, or ohms and milliamps. The equations will not work out properly with such a mismatch. To be on the safe side, it is recommended that you simply get into the habit of using only ohms and amperes in such electronic equations.

Resistor characteristics

Resistors are available in a wide variety of sizes, values, and types. Different materials can be used to make a resistor. Some typical resistor types will be discussed in a later section of this chapter.

When selecting a resistor for a specific application, three parameters are commonly considered: resistance, power rating, and tolerance. The resistance value is the most important specification for a resistor. It is given in ohms, kilohms, or megohms, depending on which unit is most convenient. Resistor values and the standard resistor color code will be introduced in the next section of this chapter.

The power rating is a measurement of how much electrical power the resistor can withstand without damage. The more power that is fed through a resistor, the greater the amount of heat that builds up within the component. If the applied power is too great, excessive heat will be generated and the resistor will burn itself up. The component could permanently change its resistance value, or it might be totally destroyed.

For a given type of resistor, the higher the power rating, the larger the physical size of the body of the component. In most practical electronic circuits, 0.25-W or 0.5-W resistors are by far the most common. Some low-power computer circuits may use tiny 0.125-W resistors. High-power circuits may require 1-W resistors or even components rated for 2 W or 5 W.

If you cannot find a resistor with the specified power ratio, you can almost always substitute a higher-rated unit. That is, a 0.5-W resistor can be substituted for a 0.25-W resistor. The only time this cannot be done is when the resistor must fit into a given space and the higher powered unit won't physically fit.

Also, high-power resistors (1 W or higher) tend to be more expensive than lighter-duty components. For the most part, there seems to be little or no difference in the cost of 0.5-W and 0.25-W resistors.

Electrical power is measured in watts, which is equal to the product of the voltage and the current. That is

$$P = EI$$

where

  P = power, in watts;
  E = voltage, in volts; and
  I = current, in amperes.

Sometimes it is convenient to combine the power formula with Ohm's law to derive the power rating from the resistance and current values:

$$P = EI$$
$$E = IR$$
$$P = (IR)I$$
$$= I \times I \times R$$
$$= I^2R$$

Similarly, if the voltage and the resistance are known, the power can be found directly, without bothering with the current value:

$$P = EI$$
$$E = IR$$
$$P = E\left(\frac{E}{R}\right)$$
$$= \frac{(E \times E)}{R}$$
$$= \frac{E^2}{R}$$

For example, let's say we have an electronic circuit powered by a 9-V battery. What is the minimum power rating for a 1-kΩ (1,000-Ω) resistor used in the main power line of this circuit?

$$P = \frac{9^2}{1000}$$
$$= \frac{(9 \times 9)}{1000}$$

$$= \frac{81}{1000}$$

$$= 0.081 \text{ W}$$

A 0.125-W resistor will do just fine, but a 0.25-W unit will provide more "headroom" and better insurance. Besides, a 0.25-W resistor is usually easier to find and less expensive than a 0.125-W resistor.

But be careful. What happens if we reduce the resistance to 270 Ω in that same 9-V circuit?

$$P = \frac{9^2}{270}$$

$$= \frac{(9 \times 9)}{270}$$

$$= \frac{81}{270}$$

$$= 0.3 \text{ W}$$

This is more than the wattage allowed by a 0.25-W resistor. You'll have to move up to a resistor rated for 0.5 W. As you can see, lower resistance values generally require larger power-handling capability, all other things being equal.

Another important specification for a resistor is tolerance. Because of manufacturing limitations, it is difficult to make every resistor come out with exactly the desired resistance. The tolerance rating indicates the guaranteed maximum deviation from the nominal value.

Until fairly recently, 10% tolerance resistors were the norm, but today, 5% tolerance resistors seem to be the most commonly used. You might occasionally come across a 20% tolerance resistor, but these devices are becoming increasingly rare in today's electronics marketplace. In precision applications, special low-tolerance resistors are used. These more expensive devices are available with tolerance ratings of 4%, 3%, 2%, 1%, and even 0.1%.

To understand what the tolerance rating for a resistor means, let's assume we are considering resistors with a nominal resistance value of 1,000 Ω (1 kΩ). A 10% tolerance resistor is guaranteed to have an actual value of ±10% of the nominal resistance value: 10% of 1,000 Ω is 100 Ω, so the resistor in our example might have an actual resistance as low as 900 Ω (1000 − 100), or as high as 1,100 Ω (1000 + 100). Note that these are the most extreme possibilities. The actual resistance will probably be somewhere between these two extremes. It might even be exactly 1,000 Ω, although that isn't terribly likely. Of course, the exact resistance value can be found with an ohmmeter. Table 2-1 lists some possible resistance values for a 1-kΩ resistor with a tolerance of 10%.

Table 2 – 1   Possible values for a 1000 – Ω, 10% tolerance resistor.

900 Ω	1000 Ω	1055 Ω
901 Ω	1001 Ω	1066 Ω
912 Ω	1003 Ω	1071 Ω
937 Ω	1007 Ω	1079 Ω
951 Ω	1011 Ω	1086 Ω
977 Ω	1017 Ω	1092 Ω
993 Ω	1024 Ω	1097 Ω
998 Ω	1043 Ω	1100 Ω

Similarly, a 1,000-Ω resistor with a tolerance of 5% would have a maximum deviation of ± 50 Ω. That is, this component is guaranteed to have an actual resistance of no less than 950 Ω and no greater than 1,050 Ω. A 1,000-Ω, 20% tolerance resistor, on the other hand, would be acceptable for our example if its actual value was as much as 200 Ω (20% of 1,000 Ω) off from the nominal value, in either direction.

A low-tolerance resistor is used in critical applications where the actual resistance must be very close to the nominal value. A 1,000-Ω resistor with a tolerance of 1% will be no more than ± 10 Ω from the nominal value (990 Ω to 1,010 Ω). Bear in mind that you can use a resistor with a higher tolerance rating if you check the actual resistance with an ohmmeter. Remember, even a 20% tolerance resistor may be exactly at its nominal value, but the manufacturer only promises that it will be within ± 20%. In other words, the tolerance rating of a resistor is rather like an insurance policy for the component's nominal resistance value.

When manufacturing resistors, a certain percentage of units will be off-value. For a high-tolerance resistor, the manufacturer must measure each resistor and make sure it is within acceptable limits. This is why high-tolerance resistors (sometimes called precision resistors) are always much more expensive than standard 5% or 10% tolerance resistors, which do not require such tight quality control during manufacturing.

Few factories deliberately manufacture 20% tolerance resistors. Usually, these are the rejects from 5% or 10% production lines. Therefore, a 20% tolerance resistor is generally not likely to be very close to its nominal value, although it is possible.

The resistor color code

There is no way to determine a resistor's value from its size or shape. The resistance value could be stamped or painted on the body of the component, but the numbers would be quite small and difficult to read. On older components, the markings could eventually rub off or become smeared. Also, when a resistor is used in a circuit, the markings would not always be visible unless the component was mounted in a very specific way. Because of these problems, a standardized color-coding system was devised for resistors. This color code is virtually universal in electronics work all over the world.

Rather than having printed numbers or letters on the body of the resistor, it is encircled with three or four colored bands. Because these bands go all the way around the body of the resistor, they are plainly visible from any angle. They are also large enough and distinctive enough to be read accurately even at a moderate distance. The bands are unlikely to rub off or be smeared. The colored bands can be very inexpensively applied to the individual resistors during manufacture.

Each band position and the color of each band are assigned specific, unambiguous meanings. A typical color-coded resistor is shown in Fig. 2-2. The resistor color code is summarized in Table 2-2.

The first two bands on the resistor are the two most significant digits of the resistance value. Ten different colors are used

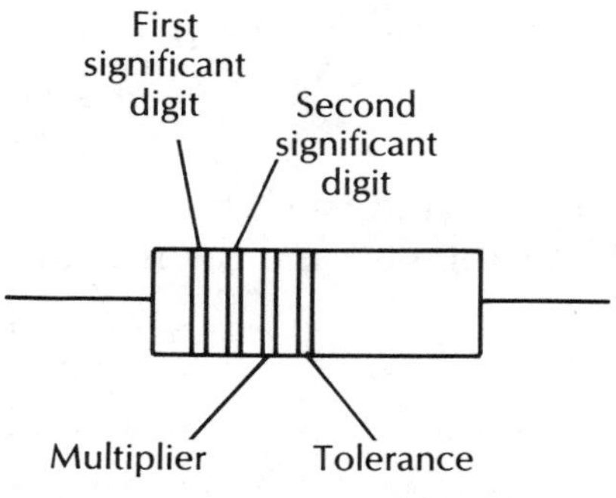

Fig. 2-2 *Resistors are usually marked with a standard color code.*

First significant digit

Second significant digit

Multiplier

Tolerance

Table 2–2   Standard color code for resistors.

Color	Band 1	Band 2	Band 3	Band 4
Black	0	0	1	—
Brown	1	1	10	1%
Red	2	2	100	2%
Orange	3	3	1,000	3%
Yellow	4	4	10,000	4%
Green	5	5	100,000	—
Blue	6	6	1,000,000	—
Violet	7	7	10,000,000	—
Gray	8	8	100,000,000	—
White	9	9	—	—
Gold	—	—	0.1	5%
Silver	—	—	0.01	10%
No color	—	—	—	20%

on these bands indicating the digits from 0 to 9. The first band is the most significant digit, and the second band is the second most significant digit.

The third band indicates a multiplier factor. This is a power of 10, and is multiplied by the two significant digits. That is, the multiplier value determines how many zeros will follow the significant digits.

Let's say, for example, the first three bands of a resistor have the following colors: yellow, violet, and red. The first band is yellow, so the most significant digit is 4. The second band is violet, indicating the second most significant digit is 7. The base value of this resistor is 47. The third band is red, so the multiplier is 100. This means the resistor's value is equal to

$$R = 47 \times 100$$
$$= 4,700 \ \Omega$$
$$= 4.7 \ k\Omega$$

If the third band was green instead of red, the multiplier factor would be 100,000 and the resistor's value would be

$$R = 47 \times 100,000$$
$$= 4,700,000 \ \Omega$$
$$= 4,700 \ k\Omega$$
$$= 4.7 \ M\Omega$$

On the other hand, very small resistances can be indicated if the multiplier (third) band is gold or silver. For example, if we again have a yellow first band and a violet second band, but a silver third band, the indicated resistance value is

$$R = 47 \times 0.01$$
$$= 0.47 \ \Omega$$

There are 99 possible combinations for the first two bands (not counting 00, which would be ridiculous and useless), but only certain base values are normally used with practical resistors. These standard resistance values are listed in Table 2-3. Each value can be repeated for each multiplier factor.

Table 2–3   Standard values for resistors (significant digits only).

10	Brown-black
12	Brown-red
(15)	Brown-green
18	Brown-gray
22	Red-red
(24)	Red-yellow
27	Red-violet
33	Orange-orange
(36)	Orange-blue
39	Orange-white
42	Yellow-red
47	Yellow-violet
(51)	Green-brown
56	Green-blue
62	Blue-red
68	Blue-gray
(75)	Violet-green
82	Gray-red
(91)	White-brown

(Values in parentheses are less commonly used.)

The fourth band (if any) on a resistor indicates the component's tolerance rating. Usually this band will be gold (5%) or silver (10%). If there is no fourth band, the resistor has a tolerance rating of 20%.

High-precision (very low tolerance—less than 1%) resistors usually have their values stamped directly on the body of the component and the standard color code is not used. Such precision resistors usually have three significant digits, which could not be indicated using the standard color code.

The resistor color code might seem a bit confusing and rather inconvenient at first. Most beginners in electronics have to keep a color code chart close at hand and refer to it very frequently. Most electronics hobbyists and technicians are surprised when they realize how quickly they've memorized the system, even without trying to do so. The resistor color code seems to be quite intuitive, and after a brief familiarization period, it is very easy to use. Experienced electronics hobbyists and technicians find this color code almost second nature and at least as easy to use as printed numbers.

Combining resistances

Occasionally, you may find you need a resistor with a nonstandard value. Resistances can be combined in both series and parallel arrangements to affect their values. Understanding how this is done is also vital for analyzing electronic circuits of all types. Every electronic component exhibits some amount of resistance. All of the resistances in a circuit must be combined to understand how the circuit functions.

Two or more resistors can be connected in series, as illustrated in Fig. 2-3. In a series combination, the resistances are added together. That is

$$R_t = R_1 + R_2 + \ldots + R_n$$

Of course, all of these resistances must be in the same unit form. That is, all values in ohms, or all values in kilohms, or all values in megohms. Don't mix units of measurement.

In the following examples, we will assume that all resistors are 100 Ω, unless otherwise noted. If we connect two 100-Ω resis-

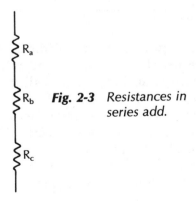

Fig. 2-3   Resistances in series add.

tors in series, we get a total effective resistance of

$$R_t = R_1 + R_2$$
$$= 100 + 100$$
$$= 200 \ \Omega$$

If there are three 100-Ω resistors in series, the total effective resistance works out to

$$R_t = R_1 + R_2 + R_3$$
$$= 100 + 100 + 100$$
$$= 300 \ \Omega$$

This formula can easily be extended to include any number of resistances in series. Notice that when resistors are connected in series, the resulting combined resistance is always greater than any of the component resistances.

Resistances can also be connected in parallel, as illustrated in Fig. 2-4. In this case, the formula for the total effective resistance is somewhat more complex. The reciprocal of the total combined resistance is equal to the reciprocals of the individual

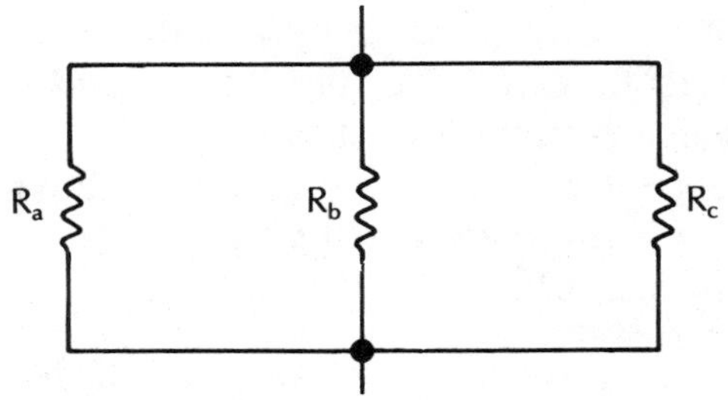

Fig. 2-4 *Resistances can also be combined in parallel.*

resistances in the parallel combination. This is easier to explain in mathematical form:

$$\frac{1}{R_t} = \frac{1}{R_1} + \frac{1}{R_2} + \frac{1}{R_3} + \ldots + \frac{1}{R_n}$$

For example, if we connect two 100-Ω resistors in parallel, the combined resistance value works out to

$$\frac{1}{R_t} = \frac{1}{R_1} + \frac{1}{R_2}$$
$$= \frac{1}{100} + \frac{1}{100}$$
$$= 0.01 + 0.01$$

$$\frac{1}{R_t} = 0.02$$

$$R_t = \frac{1}{0.02}$$

$$= 50 \; \Omega$$

Similarly, if we have three 100-Ω resistors wired in parallel, the total effective resistance will be

$$\frac{1}{R_t} = \frac{1}{R_1} + \frac{1}{R_2} + \frac{1}{R_3}$$

$$= \frac{1}{100} + \frac{1}{100} + \frac{1}{100}$$

$$= 0.01 + 0.01 + 0.01$$

$$\frac{1}{R_t} = 0.03$$

$$R_t = \frac{1}{0.03}$$

$$= 33.33 \; \Omega$$

Once again, this formula can be extended to allow for as many resistances as you have connected in parallel. Notice that when resistors are connected in parallel, the resulting combined resistance is always less than any of the component resistances.

If there are just two resistors connected in parallel, a different formula can be used:

$$R_t = \frac{(R_1 \times R_2)}{(R_1 + R_2)}$$

This modified formula is sometimes easier to work with than the usual parallel resistance equation. There is no need to take the reciprocal of the result.

For example, if we have two 100-Ω resistors connected in parallel, this alternate formula gives us a total effective resistance value of

$$R_t = \frac{(R_1 \times R_2)}{(R_1 + R_2)}$$

$$= \frac{(100 \times 100)}{(100 + 100)}$$

$$= \frac{10,000}{200}$$

$$= \frac{100}{2}$$

$$= 50 \ \Omega$$

Notice this is exactly the same result we got with the original parallel resistance equation.

This alternate parallel resistance formula can only be used with two resistances in parallel. It cannot be extended to accommodate three or more parallel resistances.

In many practical electronic circuits, you will encounter both series and parallel resistances, often in combination with one another. As an example, consider the nine-resistor network shown in Fig. 2-5. As you can see, this circuit includes both series and parallel resistance combinations.

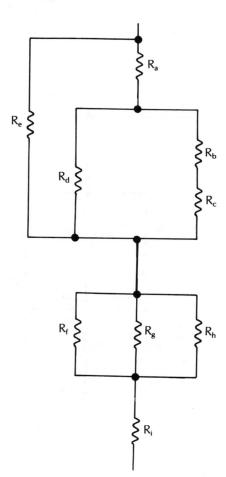

Fig. 2-5 *Practical electronic circuits usually include both series and parallel resistances.*

For our first example, we will assume that all nine resistors in this network are 100-Ω units. The first step is to find the series combination of resistors R_b and R_c:

$$R_{bc} = R_b + R_c$$
$$= 100 + 100$$
$$= 200 \ \Omega$$

Resistors R_b and R_c can be considered a single 200-Ω resistor, as shown in the revised circuit of Fig. 2-6. Now, we have to solve the parallel combination of R_d and R_{bc}:

$$R_{bcd} = \frac{(R_d \times R_{bc})}{(R_d + R_{bc})}$$
$$= \frac{(100 \times 200)}{(100 + 200)}$$
$$= \frac{20,000}{300}$$
$$= 66.66 \ \Omega$$

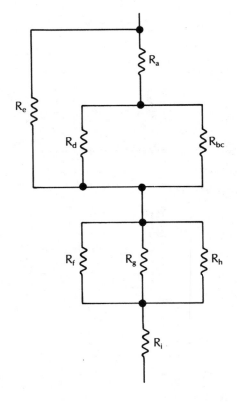

Fig. 2-6 *Simplification of the circuit in Fig. 2-5.*

We will round this off to 67 Ω.

Parallel resistance R_{bcd} can be redrawn as a single resistor, as in Fig. 2-7. This combined resistance is in series with resistor R_a, making an effective total of

$$
\begin{aligned}
R_{abcd} &= R_a + R_{bcd} \\
&= 100 + 67 \\
&= 167 \ \Omega
\end{aligned}
$$

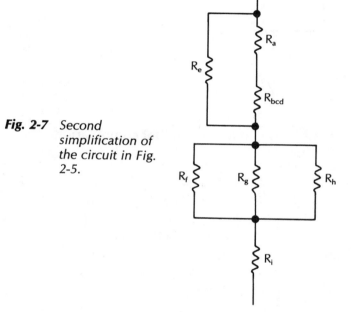

Fig. 2-7 Second simplification of the circuit in Fig. 2-5.

This resistance is in parallel with resistor R_e, for a combined total resistance of

$$
\begin{aligned}
R_{abcde} &= \frac{(R_{abcd} \times R_e)}{(R_{abcd} + R_e)} \\
&= \frac{(167 \times 100)}{(167 + 100)} \\
&= \frac{16,700}{267} \\
&= 62.55 \ \Omega
\end{aligned}
$$

We will round this off to 63 Ω, and show the combined effective resistance as a single resistor in Fig. 2-8.

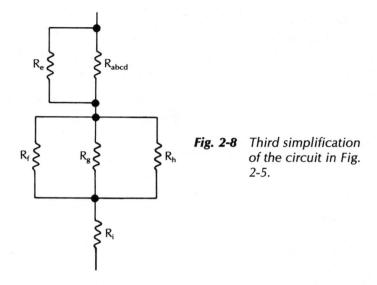

Fig. 2-8 *Third simplification of the circuit in Fig. 2-5.*

Next, we move on to the parallel combination of resistors R_f, R_g, and R_h:

$$\frac{1}{R_{fgh}} = \frac{1}{R_f} + \frac{1}{R_g} + \frac{1}{R_h}$$

$$= \frac{1}{100} + \frac{1}{100} + \frac{1}{100}$$

$$= 0.01 + 0.01 + 0.01$$

$$\frac{1}{R_{fgh}} = 0.03$$

$$R_{fgh} = \frac{1}{0.03}$$

$$= 33.33 \ \Omega$$

We will round this off to 33 Ω and redraw the circuit diagram to show this parallel combination as a single resistance unit. This is shown in Fig. 2-9.

We now have three resistances connected in series—R_{abcde}, R_{fgh}, and R_i, so the total resistance of this network is

$$R_t = R_{abcde} + R_{fgh} + R_i$$

$$= 63 + 33 + 100$$

$$= 196 \ \Omega$$

Obviously, this particular resistance network isn't too useful when all the resistors are the same value. We could just connect

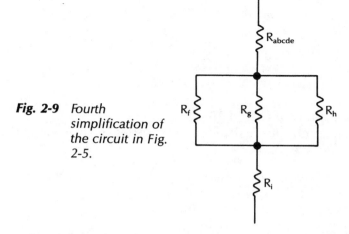

Fig. 2-9 *Fourth simplification of the circuit in Fig. 2-5.*

two of the resistors in series and achieve practically the same effect. However, no law says all the resistor values have to be equal.

Let's work through this resistor network again, but this time we will assume that we are using the varied resistance values listed in Table 2-4. Once again, the first step is to find the series combination of resistors R_b and R_c:

$$R_{bc} = R_b + R_c$$
$$= 6800 + 1000$$
$$= 7800 \ \Omega$$

Table 2-4   Resistor values used in the series/parallel problem described in the text.

R_a	3.9 kΩ
R_b	6.8 kΩ
R_c	1 kΩ
R_d	4.7 kΩ
R_e	2.2 kΩ
R_f	10 kΩ
R_g	680Ω
R_h	3.3 kΩ
R_i	1.8 kΩ

Next, we have to solve the parallel combination of R_d and R_{bc}:

$$R_{bcd} = \frac{(R_d \times R_{bc})}{(R_d + R_{bc})}$$

$$= \frac{(4700 \times 7800)}{(4700 + 7800)}$$

$$= \frac{36,660,000}{12,500}$$

$$= 2932.8 \ \Omega$$

We will round this off to 2930 Ω.

This combined resistance is in series with resistor R_a, making an effective total of

$$R_{abcd} = R_a + R_{bcd}$$
$$= 3900 + 2930$$
$$= 6830 \ \Omega$$

This resistance is in parallel with resistor R_e, for a combined total resistance of

$$R_{abcde} = \frac{(R_{abcd} \times R_e)}{(R_{abcd} + R_e)}$$

$$= \frac{(6830 \times 2200)}{(6830 + 2200)}$$

$$= \frac{15,026,000}{9030}$$

$$= 1664 \ \Omega$$

We will round this off to 1660 Ω.

The next step is to solve for the parallel combination of resistors R_f, R_g, and R_h:

$$\frac{1}{R_{fgh}} = \frac{1}{R_f} + \frac{1}{R_g} + \frac{1}{R_h}$$

$$= \frac{1}{10000} + \frac{1}{680} + \frac{1}{3300}$$

$$= 0.0001 + 0.00147 + 0.003$$

$$\frac{1}{R_{fgh}} = 0.0018735$$

$$R_{fgh} = \frac{1}{0.0018735}$$

$$= 533.7 \ \Omega$$

We will round this off to 530 Ω, and redraw the circuit as shown in Fig. 2-10.

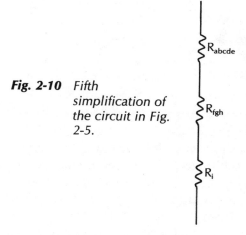

Fig. 2-10 *Fifth simplification of the circuit in Fig. 2-5.*

We now have three resistances connected in series—R_{abcde}, R_{fgh}, and R_i, and the total effective resistance of this network is:

$$
\begin{aligned}
R_t &= R_{abcde} + R_{fgh} + R_i \\
&= 1660 + 530 + 1800 \\
&= 3{,}990 \ \Omega \\
&= 3.99 \ k\Omega
\end{aligned}
$$

It is very important to realize that resistor tolerances add in both series and parallel combinations. That is, if we combine two 10% resistors in either series or in parallel, the combined effective resistance could be off from its nominal (calculated) value by as much as 20%.

Carbon resistors

Until fairly recently, the most common type of resistor has been the carbon resistor. This type of component is inexpensive, reliable, and available in a wide variety of values ranging from a fraction of an ohm up to tens of megohms (millions of ohms). Most carbon resistors have a tolerance of 5% or 10%. Rejected (out-of-tolerance) devices are rated for 20% tolerance. Some precision carbon resistors with tolerances as low as 1% (or possibly even less) are also available, although they tend to be significantly more expensive than the standard 5% or 10% type.

In appearance, a carbon resistor is a small cylinder, usually of a dark brown color. Metal leads extend from the body of the resistor on either side, as illustrated in Fig. 2-11. If you break

Fig. 2-11 *A carbon resistor is a small cylinder, usually of a dark brown color.*

open a carbon resistor, you will find it is filled with a substance that looks very much like the lead of a pencil. In fact, the active ingredient is carbon graphite, which is the same material used in pencil leads. Carbon graphite is a poor conductor, but it isn't quite an insulator. That is, this substance has a moderately large resistance. The carbon graphite in a carbon resistor is in powdered form, and this powder is mixed together with a nonconductive binder. The mix ratio of these two materials determines the actual resistance of the component. There is no physical size difference among carbon resistors of different resistances. A 0.1-Ω resistor is exactly the same size as a 10-MΩ (10,000,000-Ω) resistor.

A carbon resistor is a fairly stable component, and usually you can rely on the marked value (within the limits of the component's tolerance rating, of course). However, a resistor can change its value under certain conditions. For example, a small crack in the body of a resistor can cause the resistance to increase. Often this increase will be intermittent. That is, the resistor will irregularly change its value during operation of the circuit.

Carbon resistors have a negative temperature coefficient. This means as the temperature of the component is increased, the resistance will decrease. Usually this will not be a severe out-of-tolerance change in value unless the component is operated under relatively extreme temperature conditions. If a carbon resistor is subjected to too much heat, it can be burnt out. This will cause it to jump up to a very high resistance value or possibly become completely open. A resistor can sometimes self-destruct if it tries to carry excessive power levels. Remember, a resistor converts electrical energy into heat which must then be dissipated.

Often a burnt-out resistor can be spotted visually because the body of the component can become discolored. It will tend to be darker than normal. If the damage has happened fairly recently, a technician with a good nose can often sniff out the damaged resistor.

Most carbon resistors are 0.25-W or 0.5-W units, but 0.125-W, 1-W, and 2-W carbon resistors are not entirely uncommon. Generally, when a resistor needs to handle more than 2 W, a wire-wound resistor will be used instead of a carbon resistor. This type of component will be discussed shortly.

The higher the wattage rating of a carbon resistor, the larger its physical size will be. Power-handling capability is normally the only factor in determining the size of a carbon resistor. The resistance of the component has no effect on its size. This is because a standard-sized body is wrapped around the resistive core.

Metal-film resistors

Today, carbon resistors are being replaced by metal-film resistors. Instead of a carbon graphite powder, this type of resistor uses a thin metal film wrapped around an insulating core.

While it isn't always a 100% sure indication, metal-film resistors are often slightly smaller than a comparable carbon resistor. Instead of being a true cylinder, a typical metal-film resistor usually has a rather bonelike shape, as illustrated in Fig. 2-12.

Fig. 2-12 *A metal-film resistor often has a bonelike shape.*

The metal-film technique is an advantage to manufacturers of resistors, but it generally doesn't mean much to electronics hobbyists and technicians. With very few exceptions, carbon resistors and metal-film resistors can be considered interchangeable. There is one slight advantage to metal-film resistors. They tend to have a slightly lower temperature coefficient than carbon resistors. That is, a given increase in temperature will cause a carbon resistor to drop more in resistance than a comparable metal-film resistor.

Wire-wound resistors

Another fairly common type of resistor is the wire-wound resistor. Wire-wound resistors are most useful in high-power circuits, particularly when large currents are involved. A wire-wound resistor can safely dissipate greater amounts of heat than carbon resistors or metal-film resistors. Some wire-wound resistors can handle several hundred watts of continually applied power.

A wire-wound resistor is made from a length of resistive wire, such as nichrome, that is wrapped around a nonconductive core. A ceramic core is usually, but not always, used in this type of component. The resistance of a wire-wound resistor depends on the characteristics of the particular resistive wire used and the wire's length. Obviously a long (higher resistance) wire will make more turns around the core. The size of the resistive wire also contributes to determining the power-handling capability of the component. The diameter of the core and the material it is made out of also affect the maximum power dissipation of the resistor.

A wire-wound resistor is designed to provide a dc resistance, but the nature of its construction also makes it act like a coil (see chapter 4). In addition to the desired dc resistance, a wire-wound resistor exhibits inductive reactance (ac resistance). This characteristic makes wire-wound resistors unsuitable for use in most radio frequency (rf) circuits.

As a rule, wire-wound resistors are considerably more expensive and bulkier than carbon or metal-film resistors, but they can handle larger amounts of current without burning out.

Rheostats

In many circuits, a manually variable resistance is desirable or essential. There are two basic types of manually variable resistance devices—rheostats and potentiometers.

Basically, a rheostat is a wire-wound resistor with a movable slider that can make contact with the resistance wire at any point along its length. This is illustrated in Fig. 2-13. For a given size and type of resistance wire, the resistance depends on the length of the wire. In a fixed wire-wound resistor, the length is obviously fixed during manufacture.

In a rheostat, one end of the resistance wire serves as one lead of the variable resistor. The other lead is the slider, which can be moved up and down the length of the resistance wire. This varies the effective length of the wire between the two leads and, thus, the resistance.

Some, but not all rheostats, have a third lead at the opposite end of the coiled resistance wire. The wire length, and therefore

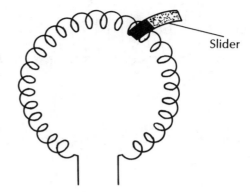

Slider

Fig. 2-13 A rheostat is a
wire-wound
resistor with a
movable slider.

the resistance, between these two end leads is constant. We will
call these two end leads A and B.

With three leads (counting the slider, or S), there are three
resistances that can be used. Two of these resistances are variable, but complementary. The third is a fixed resistance. These
three resistances of the rheostat can be identified as

- R_{AB}—fixed resistance between A and B;
- R_{AS}—variable resistance between A and the slider; and
- R_{BS}—variable resistance between B and the slider.

The two variable resistances always add up to the fixed resistance. That is

$$R_{AB} = R_{AS} + R_{BS}$$

In other words, as resistance R_{AS} increases, R_{BS} decreases, and
vice versa. Resistance R_{AB}, of course, has a constant, unchanging
value.

If a rheostat is in the form of a straight cylinder, as shown in
Fig. 2-14, it is called a solenoidal rheostat. A rheostat may also be
designed in a circular shape, as illustrated in Fig. 2-15. This type
of rheostat is known as a rotary rheostat or a toroidal rheostat.

There is no difference in the way these two types of rheostats
function. The only differences are the shape and the physical
movement of the slider. On a solenoidal rheostat, the slider moves
back and forth along a straight line. A toroidal rheostat, on the
other hand, uses a rotating shaft as the slider. Sometimes a screwdriver adjustment is used to position the slider on a rheostat.

A rheostat can withstand fairly high power levels and large
current flows. Like a wire-wound resistor, a rheostat also acts like

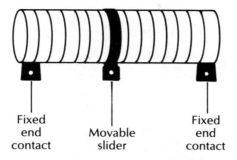

Fig. 2-14 *Some rheostats are solenoidal rheostats.*

Fixed end contact   Movable slider   Fixed end contact

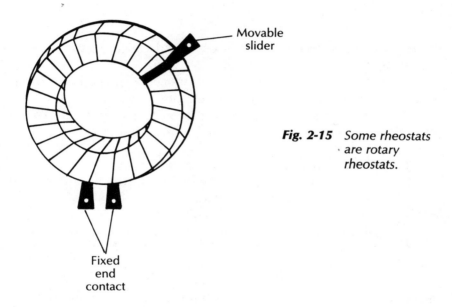

Movable slider

Fig. 2-15 *Some rheostats are rotary rheostats.*

Fixed end contact

a coil. In addition to the desired dc resistance, a rheostat exhibits inductive resistance, which could be highly undesirable in ac circuits utilizing high-frequency signals, such as rf circuits.

The standard symbol for a two-lead rheostat is shown in Fig. 2-16. One end of the zigzag resistor line is the fixed lead, and the other end is the slider. It doesn't matter which is which, because a rheostat is not a polarized device. Notice that this is basically the standard resistor symbol with an arrow through it to indicate that the resistance is variable.

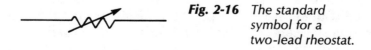

Fig. 2-16 *The standard symbol for a two-lead rheostat.*

A three-lead rheostat is usually indicated by the symbol shown in Fig. 2-17. The two ends of the zigzag resistor line are the fixed end terminals of the rheostat. The slider is represented by the center arrow.

Fig. 2-17 The standard symbol for a three-lead rheostat.

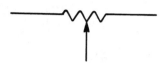

A three-lead rheostat can easily be used in place of a two-lead rheostat. Either leave one of the fixed-end connections unattached, or short one end terminal to the slider, as illustrated in Fig. 2-18.

Fig. 2-18 A three-lead rheostat can easily be made to simulate a two-lead rheostat.

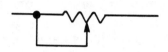

Potentiometers

A potentiometer is similar to a rheostat. It is a manually variable resistance device. While the rheostat is a variable version of the fixed wire-wound resistor, the potentiometer is a variable version of the fixed carbon resistor.

A potentiometer, like a fixed carbon resistor, usually can't withstand the high power levels that can be handled by a rheostat or a fixed wire-wound resistor. On the other hand, a potentiometer, like a fixed carbon resistor, exhibits dc resistance. There is no ac inductive reactance, as in a rheostat or a fixed wire-wound resistor.

Instead of moving along a coiled wire, the slider in a potentiometer moves along an arc of granulated carbon graphite, as illustrated in Fig. 2-19. The slider is connected to a rotary shaft. In use, a knob is usually mounted on the end of the shaft. Sometimes a graduated dial may also be used to make setting a specific resistance more convenient.

Occasionally, two, or sometimes three or more, individual potentiometers are ganged together. A single shaft is used to simultaneously control the sliders of all of the ganged potentiometers. Some potentiometers also feature a ganged on-off switch.

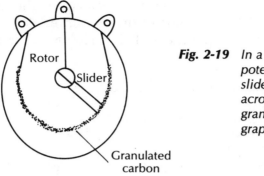

Fig. 2-19 *In a potentiometer, a slider is moved across a strip of granulated carbon graphite.*

Rotor

Slider

Granulated carbon

When the shaft is turned to an extreme resistance position (either minimum or maximum, depending on the application), the switch clicks off. Moving the shaft from this extreme position turns the switch on. A potentiometer is sometimes referred to as a pot.

When selecting a potentiometer, the taper is often important. The taper of a potentiometer is a description of how the resistance varies with the movement of the control shaft. The two most common types of potentiometers are linear taper and logarithmic taper. Both of these terms refer to how the resistance-to-control shaft position can be graphed.

A linear taper potentiometer varies its resistance linearly with the movement of the control shaft, as illustrated in Fig. 2-20. If rotating the shaft x degrees changes the resistance by a

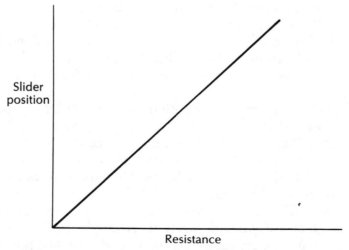

Slider position

Resistance

Fig. 2-20 *Some potentiometers have a linear taper.*

factor of y, then rotating the shaft 2x degrees will change the resistance by a factor of 2y.

A graph of a logarithmic taper potentiometer is shown in Fig. 2-21. A given amount of rotation of the control shaft at one end of its travel will result in a very different change in resistance as an equal amount of rotation at the opposite end of the potentiometer's range. The logarithmic taper potentiometer is often called an audio taper potentiometer. This type of potentiometer is commonly used as a volume control in audio amplifiers. The reason for this is that the human ear hears changes in volume in a logarithmic rather than a linear fashion.

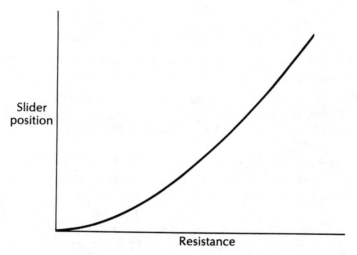

Fig. 2-21 *Some potentiometers have a logarithmic taper.*

A few potentiometers use different, more exotic tapers, such as reverse audio taper potentiometers. The action of this device is illustrated in Fig. 2-22.

Most standard potentiometers use rotary shafts, but there are also some carbon potentiometers that more closely resemble the solenoidal rheostat. That is, the slider is moved back and forth along a straight line rather than in a rotary fashion. This device is commonly known as a slide pot. The strip of granulated carbon graphite is simply placed along a straight line rather than in a curved arc, as in an ordinary potentiometer. Other than the difference in the mechanical motion, a slide pot works in the same way as a standard rotary potentiometer. Most slide pots use a linear taper. Slide pots are commonly used in applications where

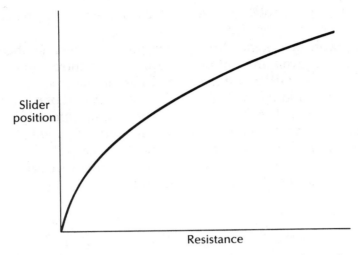

Fig. 2-22 *A few potentiometers have exotic tapers, such as the reverse audio taper.*

the shaft position must be directly visible for some reason. Slide pots are widely used in audio equipment, such as mixers and graphic equalizers.

Miniature potentiometers are also available. The slider in such a device is usually controlled with a screwdriver adjustment. This device is known as a trimmer potentiometer or, more commonly, as a trimpot. Trimpots are normally employed for calibration controls. Once they are set, they are usually not intended to be readjusted during normal use. Most common trimpots are rated for no more than 0.25 W, although some 0.5-W trimpots have been manufactured. Most standard (full-sized) potentiometers can handle power levels up to 2 W.

With a few exceptions, most standard potentiometers have three leads, counting the slider. The two end terminals are fixed. We will call them A and B. We will identify the movable slider terminal as S.

With these three terminals, three resistances can be used. Two of these resistances are variable, but complementary. The third is a fixed resistance. These three resistances of the potentiometer can be identified as:

- R_{AB}—fixed resistance between A and B;
- R_{AS}—variable resistance between A and the slider; and
- R_{BS}—variable resistance between B and the slider.

The two variable resistors always add up to the fixed resistance. That is:

$$R_{AB} = R_{AS} + R_{BS}$$

In other words, as resistance R_{AS} increases, R_{BS} decreases, and vice versa. Resistance R_{AB}, of course, has a constant, unchanging value. Neither R_{AS} nor R_{BS} can ever exceed the value of R_{AB}. This is the maximum resistance of the potentiometer. The R_{AB} fixed resistance is listed as the value of the potentiometer. For example, a 50-kΩ potentiometer has a resistance of 50,000 Ω between its end (fixed) terminals (A and B). The slider can be adjusted to create an adjustable resistance ranging between a little over 0 Ω to a little under 50,000 Ω.

Most standard potentiometers have three terminals. A few trimpots have just two terminals. In this case, one of the end terminals is simply omitted.

The same standard schematic symbols are commonly used to represent both rheostats and potentiometers. The current flow through the circuit (and possibly the signal frequency) will determine which device is to be used. When in doubt, check the parts list. In almost all low-power circuits (especially electronics hobbyist projects), potentiometers are used almost exclusively. These days, rheostats are restricted mostly to use in high-power industrial circuits.

A standard three-lead potentiometer is usually indicated by the symbol shown in Fig. 2-23. Notice that this symbol is a variation on the basic symbol used to represent a standard fixed resistor. The two ends of the zigzag resistor line in the symbol are the fixed end terminals of the rheostat. The slider is represented by the center arrow.

Fig. 2-23 *The standard symbol for a three-lead potentiometer.*

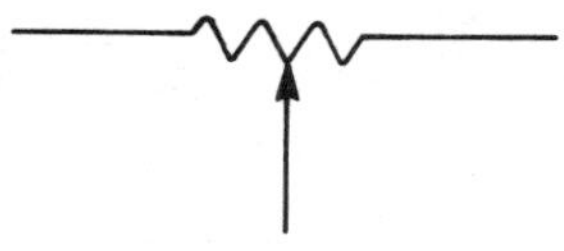

A two-lead trimpot can be represented by the standard three-lead symbol with one end left unconnected, or by the symbol shown in Fig. 2-24. In this symbol, one end of the zigzag resistor line is the fixed lead and the other end is the slider. It doesn't matter which is which because a potentiometer is not a polarized device.

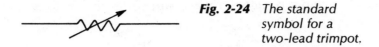

Fig. 2-24 *The standard symbol for a two-lead trimpot.*

Naturally, a standard three-lead potentiometer can be substituted for a two-lead unit. All you have to do to make this substitution is to leave one end of the fixed end connections unattached to the circuit. Alternately (and some technicians believe this method is preferable), you can short one of the fixed end terminals to the slider terminal.

❖ 3
Capacitors

ANOTHER IMPORTANT PASSIVE ELECTRONIC COMPONENT IS THE capacitor. In most electronic circuits, capacitors are almost as common as resistors. A standard resistor is intended to work in pretty much the same way on both dc and ac signals. A capacitor, on the other hand, tends to block dc signals but passes ac signals. The ac resistance (or, more properly, reactance) decreases as the signal frequency increases.

Capacitors are sometimes referred to as condensers, especially in older technical literature. These terms refer to the exact same component. Modern accepted terminology for this component is capacitor. The other term is considered archaic.

Several different symbols are used to represent capacitors in schematic diagrams. They are all fairly obvious variations on one another. The most commonly used symbol for a capacitor is the one shown in Fig. 3-1, but the symbols shown in Fig. 3-2 are not uncommon. There is no difference in meaning between these three symbols.

Most capacitors are nonpolarized components. That is, it doesn't matter which plate is connected to the negative terminal of the voltage source and which is connected to the positive terminal. (The plates of a capacitor and their meaning and significance are explained in the next section of this chapter.) There is no way to hook up a nonpolarized component backwards. Some specialized types of capacitors (capacitor types will be discussed later in this chapter) are polarized, however. A polarized capacitor must not be placed in a circuit backwards. To avoid this, the

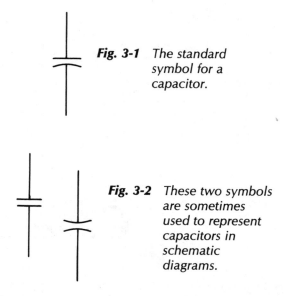

Fig. 3-1   *The standard symbol for a capacitor.*

Fig. 3-2   *These two symbols are sometimes used to represent capacitors in schematic diagrams.*

polarity must be indicated in the schematic diagram. Therefore, the standard capacitor symbol can be modified slightly, as shown in Fig. 3-3. As you can see, all we have done is add a small plus sign (+) to indicate the positive plate of the capacitor. The convention is to make the flat plate in the schematic symbol the positive plate. The symbols shown in Fig. 3-2 usually aren't used to represent polarized capacitors.

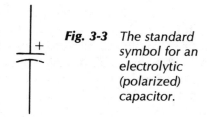

Fig. 3-3   *The standard symbol for an electrolytic (polarized) capacitor.*

How capacitance works

There are many different types of capacitors, but they all have certain characteristics in common. Essentially, a capacitor consists of two metal (conductive) plates separated by a piece of insulating material. The insulator separating the two metal plates in a capacitor is called the dielectric.

If a dc voltage is applied across the two plates, as shown in Fig. 3-4, current will not be able to cross the insulating dielec-

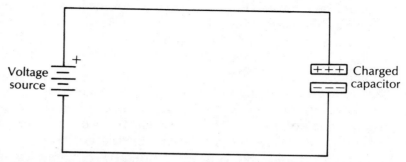

Fig. 3-4 *A dc voltage applied to the two plates of a capacitor cannot cross the dielectric.*

tric. A surplus of electrons will build up on the plate connected to the negative terminal of the voltage source. Meanwhile, there will be a shortage of electrons (positive electrical charge) on the opposite plate, which is connected to the positive terminal of the voltage source. Basically, the dc voltage source is trying to force electrons into one plate (negative terminal) and draw electrons out of the other plate (positive terminal).

At some point, the plates in the capacitor will be completely saturated. No more electrons can be forced into the negative plate, and no more electrons can be drawn out of the positive plate. At this point, the capacitor's plates have an electrical potential equal to that of the voltage source. In fact, the capacitor plates effectively act like a second voltage source in parallel with the first, but with the opposite polarity. Figure 3-5 shows the equivalent circuit for this condition. Naturally, since these opposing voltages are equal, they cancel each other out, and no

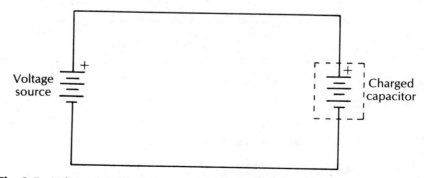

Fig. 3-5 *When charged, a capacitor effectively looks like a second voltage source in parallel with the circuit's actual voltage source.*

current can flow between the (true) voltage source and the capacitor plates in either direction. The capacitor is said to be charged. As long as the voltage source continues applying its dc voltage across the plates of the capacitor, this condition will remain unchanged.

Now, suppose we remove the dc voltage source from the circuit, as shown in Fig. 3-6. The capacitor plates will stay charged because there is no place for the excess electrons on the negative plate to go. Similarly, there is no place for the positive plate to draw electrons from. The voltage is effectively stored by the plates.

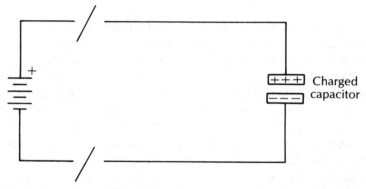

Fig. 3-6 *A capacitor will hold its charge when the dc voltage source is removed from the circuit.*

If we replace the removed voltage source with a resistor, as illustrated in Fig. 3-7, the resistor will provide a current path for the excess electrons stored on the capacitor's negative plate to flow to the positively charged plate. This current flow will con-

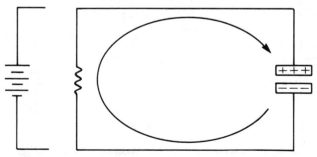

Fig. 3-7 *Replacing the voltage source with a resistor permits the capacitor to discharge.*

tinue until both plates are restored to an electrically neutral state. This process is called discharging the capacitor.

A practical capacitor cannot really hold a charge indefinitely. No insulator is 100% perfect. Even air can conduct some current, so the charge on the capacitor will slowly seep off, or dissipate. This is a form of leakage. There will also be some inevitable leakage through the capacitor's own insulating dielectric. Of course, if all other factors are equal, the lower the internal leakage, the better the capacitor.

It should be obvious that a capacitor would only find limited applications in a purely dc circuit. This type of component is more useful in circuitry carrying ac signals. Let's consider what happens within a capacitor when an alternating current is applied to it. The basic circuit is illustrated in Fig. 3-8.

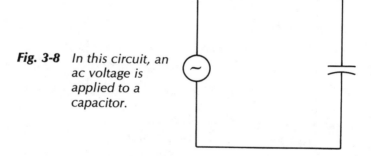

Fig. 3-8   *In this circuit, an ac voltage is applied to a capacitor.*

During the first part of the cycle, as the source voltage increases from zero, the plates of the capacitor are charged in a manner similar to the dc circuit described above. The polarity of the charged capacitor opposes the source voltage. The capacitor may or may not be fully charged by the time the source voltage passes its peak and starts to decrease again. This will depend on the size of the capacitor's plates, how much voltage is applied, and the frequency of the applied ac signal.

Whether the capacitor is fully charged or not, as the applied voltage decreases, a point will be reached when the instantaneous source voltage is less than the charge stored in the capacitor. This will allow the capacitor to start discharging through the ac voltage source. The capacitor may or may not be completely discharged when the ac voltage reverses polarity for the second half of its cycle, but because the source polarity is now the same as the capacitor charge polarity, the voltages add, quickly discharging the capacitor the rest of the way, if necessary. Once the capacitor is fully discharged, it starts being charged up with the opposite polarity from the original charge. When the ac source voltage reverses direction, the capacitor is discharged again, and the entire sequence is repeated with the next cycle of the applied ac waveform.

If you constructed the simple circuit shown in Fig. 3-9 with a dc voltage source, nothing would happen. The lamp would not light because the dc current cannot flow through the circuit. It is blocked by the insulating dielectric of the capacitor. The capacitor acts almost like an open circuit as far as direct current is concerned.

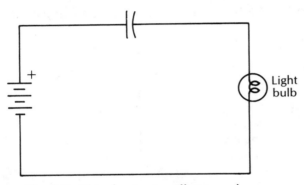

Fig. 3-9   *This dc circuit will not work.*

If, however, the same circuit is built with an ac voltage source, as shown in Fig. 3-10, the lamp will light up. This indicates that alternating current is flowing through the circuit. Of course, virtually no current (except the tiny leakage current) will actually flow across the dielectric itself. The process of charging, discharging, and recharging the capacitor with the ac voltage

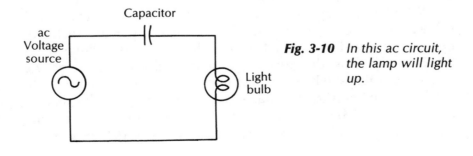

Fig. 3-10   *In this ac circuit, the lamp will light up.*

gives the same effect as if the current was actually flowing through the capacitor itself.

Moreover, if we decrease the frequency of the ac source, the lamp will dim. Increasing the applied signal frequency will cause the lamp to burn brighter. A capacitor lets more current flow as the frequency of the source voltage is increased. This property of capacitors will be fully explained a little later in this chapter.

The farad

The basic unit of capacitance is the farad (F), named for Michael Faraday (1791 – 1867), a pioneering researcher in the principles of electricity.

In an ac circuit with a capacitor, if 1 A of current flows when the applied voltage changes at a rate of 1 V/s, we have 1 F of capacitance. In practical electronic circuits, the farad is too large a unit to conveniently express reasonable capacitance values. A much smaller unit, known as a microfarad (μF) is used instead. One microfarad is equal to one-millionth of a farad:

$$1 \ \mu\text{F} \ = \ 0.000001 \ \text{F}$$

Another way to say the same thing is it takes one million micro-farads to make up one farad:

$$1 \ \text{F} \ = \ 1,000,000 \ \mu\text{F}$$

Some older technical literature, and manuals and magazines from some foreign countries (such as England), may use mF as the abbreviation for microfarads. Remember, these two labels (μF and mF) represent exactly the same thing. The terminology is interchangeable.

In some cases, the microfarad may still be too large a unit for convenient representation of the capacitance values in a circuit. A still-smaller unit in common use is the picofarad, which is equal to one-millionth of a microfarad. The standard abbreviation for the picofarad is pF. Some sources may refer to picofarads as micromicrofarads, abbreviated as $\mu\mu$F or mmF. Once again, the only difference is in the terminology used. The same units are being referred to by all three of these terms.

Several factors determine the actual capacitance of a capacitor. The most important of these factors are the size of the plates, the thickness of the dielectric (the separation of the plates), and the type of material used for the dielectric. Assuming the dielectric material is not changed, the capacitance can be increased either by making the metal plates larger or by bringing them closer together (that is, making the dielectric thinner).

Several different materials are commonly used for the dielectric of capacitors. Each insulator has a specific dielectric constant (K), which has a direct effect on the capacitance. Ordinary air, for example, has a dielectric constant of one. Mica, on the other hand, has a dielectric constant of six. Paper's dielectric constant is between two and three, and titanium oxide can have a dielectric constant as high as 170. Many other materials are used as the dielectric in practical capacitors, each with its own specific dielectric constant.

Capacitance can be found with the following formula:

$$C = \frac{(0.0885 \times K \times A)}{T}$$

where

C = capacitance, in picofarads;
K = dielectric constant;
A = area of the side of one of the plates that is actually in physical contact with the dielectric, in square centimeters; and
T = thickness of the dielectric (or the spacing between the plates), in centimeters.

As an example, let's find the capacitance of a capacitor with an air dielectric ($K = 1$). This hypothetical capacitor has the following dimensions:

- A = 35 cm^2,
- T = 2 cm.

We can now plug these values into the equation and solve for the capacitance:

$$C = \frac{(0.0885 \times K \times A)}{T}$$
$$= \frac{(0.0885 \times 1 \times 35)}{2}$$
$$= \frac{3.0975}{2}$$
$$= 1.54875 \text{ pF}$$

However, if we keep the physical dimensions of the capacitor exactly the same, but change the dielectric to mica, the dielectric constant becomes 6, so the capacitance changes to

$$C = \frac{(0.0885 \times K \times A)}{T}$$

$$= \frac{(0.0885 \times 6 \times 35)}{2}$$

$$= \frac{18.585}{2}$$

$$= 9.2925 \text{ pF}$$

Increasing the dielectric constant increases the capacitance.

Next, let's take our hypothetical mica capacitor and try increasing the value of the plate area to 50 cm². In this case, the capacitance becomes

$$C = \frac{(0.0885 \times 6 \times 50)}{2}$$

$$= \frac{26.55}{2}$$

$$= 13.275 \text{ pF}$$

Once again, increasing the plate area of a capacitor also increases its capacitance.

Finally, let's try increasing the value of T (dielectric thickness) to 5 cm. This changes the capacitance to a value of

$$C = \frac{(0.0885 \times 6 \times 50)}{5}$$

$$= \frac{26.55}{5}$$

$$= 5.31 \text{ pF}$$

Increasing the dielectric constant (K) or the plate area (A) increases the capacitance, but increasing the dielectric thickness or plate separation (T) decreases the capacitance of the device.

In actual practical electronics work, you will rarely, if ever, need to bother with these equations unless you are manufacturing capacitors. Still, it can be helpful to understand how these variables interrelate in determining the capacitance of a capacitor.

The capacitance value will usually (though not always) be marked on the body of the capacitor. If the capacitance value is not marked, it can be read with a capacitance meter. This type of test instrument is becoming increasingly common today.

Capacitive reactance

A resistor exhibits dc resistance. A capacitor has a very high dc resistance, but it also exhibits a type of ac resistance known as capacitive reactance. In electronics formulas, capacitive reactance is commonly represented as X_C.

Another type of ac resistance is inductive reactance. Inductive reactance will be discussed in chapter 4. The total ac resistance in a circuit is called the impedance, which is the combination of the dc resistance, the capacitive reactance, and the inductive reactance.

If we measure the dc resistance of a discharged capacitor with an ohmmeter, the meter's needle will show a sharp kick down to a moderately low resistance as the capacitor is charged by the ohmmeter's test voltage. Then the pointer will move up, finally settling at a very high resistance. In an ideal capacitor, the final dc resistance would be infinite. That is, the ideal capacitor would look like an open circuit to a dc voltage. A true infinite resistance is not possible with a practical real-world component, of course, but the dc resistance of a charged capacitor will be quite high. For all practical purposes, the charged capacitor is more or less an open circuit, blocking the flow of dc current.

However, as we've already seen, an ac signal can flow through a circuit with a capacitor. Some of the ac signal will be lost as it passes through the capacitor, just as some of the dc signal is lost when it passes through a resistor. This loss is the result of the capacitor's ac resistance, or capacitive reactance.

Capacitive reactance, like dc resistance, is measured in ohms (or kilohms or megohms, if appropriate). The dc resistance is the same for both dc and ac signals. The signal frequency is completely irrelevant, but capacitive reactance is frequency dependent. The value of the capacitive reactance decreases as the applied frequency increases. This is easy to remember when you consider that a dc signal has a frequency of 0 Hz and faces the highest capacitive reactance from the capacitor.

A capacitive reactance also slows down the voltage more

than it slows down the current. In a purely capacitive circuit (with no inductive reactance) the voltage will lag the current by 90 degrees, as illustrated in Fig. 3-11. That is, the voltage cycle begins one-quarter of the way through the current cycle.

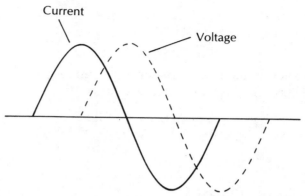

Fig. 3-11   *In a purely capacitive circuit, the voltage lags the current by 90 degrees.*

For any given capacitor, the capacitive reactance depends on the frequency of the applied signal and the capacitance value of the component. The formula for capacitive reactance is

$$X_C = \frac{1}{2\pi FC}$$

where

X_C = capacitive reactance, in ohms;
F = applied signal frequency, in hertz; and
C = capacitance, in farads.

The symbol π is the Greek letter pi, and it is used in many electronics formulas to represent a universal constant. This is not an arbitrarily selected value, but a characteristic constant that occurs repeatedly in nature. The value of π is always approximately 3.14, so 2π is equal to 6.28. This means the capacitive reactance formula can be rewritten as

$$X_C = \frac{1}{6.28FC}$$

Let's try calculating the capacitive reactance for a few practical examples. First, we'll assume that we are using a 1-μF

(0.000001-F) capacitor. If a 10-Hz signal is applied to this capacitor, the capacitive reactance will be

$$X_C = \frac{1}{(6.28 \times 0.000001 \times 10)}$$

$$= \frac{1}{0.0000628}$$

$$= 15,924 \ \Omega$$

If we take that same 1-μF capacitor and increase the signal frequency to 500 Hz, the capacitive reactance drops to

$$X_C = \frac{1}{(6.28 \times 0.000001 \times 500)}$$

$$= \frac{1}{0.00314}$$

$$= 318 \ \Omega$$

As you can see, the capacitive reactance decreases as the applied signal frequency increases.

Incidentally, this formula also predicts the correct results for a dc signal. A dc voltage has a frequency of 0 Hz. Plugging this value into our capacitive reactance formula, gives us

$$X_C = \frac{1}{(6.28 \times 0.000001 \times 0)}$$

$$= \frac{1}{0}$$

$$= \infty \ (\text{infinity})$$

The capacitive reactance is infinite when the applied signal frequency is 0 Hz (dc). The finite value measured by the ohmmeter, as mentioned above, is the dc resistance, not the capacitive reactance.

Returning to our examples, let's replace the 1-μF capacitor with a 0.22-pF (0.00000022-F) unit and find the capacitive reactance for the same two signal frequencies. When the applied signal across this capacitor has a frequency of 10 Hz, the capacitive reactance works out to

$$X_C = \frac{1}{(6.28 \times 0.00000022 \times 10)}$$

$$= \frac{1}{0.0000138}$$

$$= 72,464 \ \Omega$$

Now, if we increase the signal frequency to 500 Hz, the capacitive reactance drops to

$$X_C = \frac{1}{(6.28 \ \times \ 0.00000022 \ \times \ 500)}$$

$$= \frac{1}{0.00069}$$

$$= 1,448 \ \Omega$$

As you can see by comparing these results to our earlier examples, decreasing the capacitance for a given signal frequency increases the capacitive reactance. Some additional examples of capacitive reactance are listed in Table 3-1.

Table 3–1 Some examples of capacitive reactance for different capacitances and frequencies.

Capacitance	Frequency		
	10 Hz	**100 Hz**	**1000 Hz**
0.001 μF	16 MΩ	1.6 MΩ	160 kΩ
0.0022 μF	7.2 MΩ	720 kΩ	72 kΩ
0.0033 μF	4.8 MΩ	480 kΩ	48 kΩ
0.005 μF	3.2 MΩ	320 kΩ	32 kΩ
0.01 μF	1.6 MΩ	160 kΩ	16 kΩ
0.022 μF	720 kΩ	72 kΩ	7.2 kΩ
0.033 μF	480 kΩ	48 kΩ	4.8 kΩ
0.05 μF	320 kΩ	32 kΩ	3.2 kΩ
0.1 μF	160 kΩ	16 kΩ	1.6 kΩ
0.22 μF	72 kΩ	7.2 kΩ	720 Ω
0.33 μF	48 kΩ	4.8 kΩ	480 Ω
0.5 μF	32 kΩ	3.2 kΩ	320 Ω
1 μF	16 kΩ	1.6 kΩ	160 Ω
2.2 μF	7.2 kΩ	720 Ω	72 Ω
3.3 μF	4.8 kΩ	480 Ω	48 Ω
5 μF	3.2 kΩ	320 Ω	32 Ω
10 μF	1.6 kΩ	160 Ω	16 Ω
22 μF	720 Ω	72 Ω	7.2 Ω
33 μF	480 Ω	48 Ω	4.8 Ω
50 μF	320 Ω	32 Ω	3.2 Ω
100 μF	160 Ω	16 Ω	1.6 Ω
220 μF	72 Ω	7.2 Ω	0.72 Ω
330 μF	48 Ω	4.8 Ω	0.48 Ω
500 μF	32 Ω	3.2 Ω	0.32 Ω

Ohm's law works with reactance in exactly the same way it does with regular dc resistance; that is,

$$E = IR$$

The only difference that you must keep in mind is that the result of any specific equation involving reactance is true only for a single, specific frequency. Higher frequency signals will be subjected to a lower capacitive reactance for a given capacitor than lower frequency signals.

Combining capacitances

Like resistors, capacitors can be wired into series and parallel combinations to create new capacitance values. A circuit with two capacitors connected in parallel is shown in Fig. 3-12.

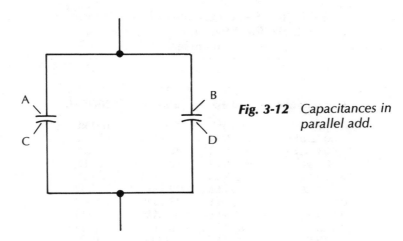

Fig. 3-12 *Capacitances in parallel add.*

Since plates A and B are tied together, they are at the same electrical potential. We can think of them as a single plate, with the area of plate A added to the area of plate B. There is no electrical difference between these two separate plates and the effective single plate of the combined dimensions. Similarly, plates C and D are electrically connected to the same potential, so they act like a single effective plate.

Remember that the larger the surface area of the plates in a capacitor, the higher the capacitance will be. Obviously, combination plate AB is going to be larger than either plate A or plate B separately. The same is true of combination plate CD.

The total effective capacitance of multiple capacitors in parallel always increases. The total effective capacitance is larger than any of the separate component capacitances. In fact, we can simply add the capacitances of capacitors in parallel to find the total effective capacitance. That is, for n capacitors in parallel

$$C_t = C_1 + C_2 + C_3 + \ldots + C_n$$

Notice that this is the same basic formula used for resistors in series.

For example, let's say we have two 0.1-μF capacitors connected in parallel. The total effective capacitance of this combination works out to

$$C_t = 0.1 + 0.1$$
$$= 0.2$$

Be sure to use like units for each component capacitance and the result. Do not combine farads, microfarads, and picofarads.

As a more complex example, let's say we have four capacitors with the following values wired in parallel:

- C_1 = 0.01 μF;
- C_2 = 0.005 μF;
- C_3 = 0.0033 μF; and
- C_4 = 0.022 μF.

The total combined capacitance value in this case works out to

$$C_t = 0.01 + 0.005 + 0.0033 + 0.022$$
$$= 0.0403 \ \mu F$$

Similarly, capacitors connected in series, as shown in Fig. 3-13, work against each other, reducing the total effective capacitance of the combination. The formula for multiple capacitances in series mirrors the formula for multiple resistances in parallel:

$$\frac{1}{C_t} = \frac{1}{C_1} + \frac{1}{C_2} + \frac{1}{C_3} + \ldots + \frac{1}{C_n}$$

Therefore, two 0.1-μF capacitors in series act like a single 0.05-

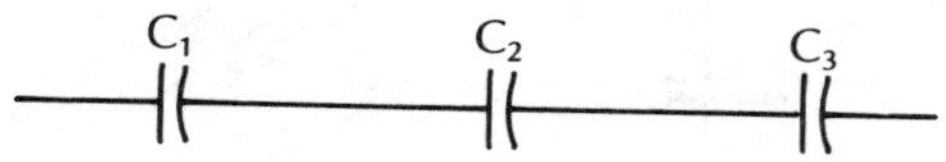

Fig. 3-13 *Capacitors can also be connected in series.*

μF capacitor. If the same two capacitors were connected in parallel, they would equal 0.2 μF.

Let's assume the four capacitance values from our previous example are connected in series, instead of parallel. In this case, the total effective capacitance works out to

$$\frac{1}{C_t} = \frac{1}{0.01} + \frac{1}{0.005} + \frac{1}{0.0033} + \frac{1}{0.022}$$

$$= 100 + 200 + 303 + 45$$

$$\frac{1}{C_t} = 648$$

$$C_t = \frac{1}{648}$$

$$= 0.00154 \ \mu F$$

Always be careful when using these formulas. Since the capacitance formulas are the mirror images of the resistance formulas, it is easy to get confused if you're not careful.

Of course, both series and parallel combinations of capacitors can be included within a single circuit, as with resistors. For example, consider the network of capacitors illustrated in Fig. 3-14. We will assume the following component values:

- C_a = 0.1 μF;
- C_b = 0.033 μF;
- C_c = 0.0015 μF; and
- C_d = 0.22 μF.

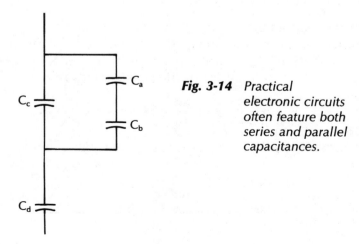

Fig. 3-14   *Practical electronic circuits often feature both series and parallel capacitances.*

The first step is to solve for the series combination of capacitors C_a and C_b:

$$\frac{1}{C_{ab}} = \frac{1}{C_a} + \frac{1}{C_b}$$

$$= \frac{1}{0.1} + \frac{1}{0.033}$$

$$= 10 + 30$$

$$\frac{1}{C_{ab}} = 40$$

$$C_{ab} = \frac{1}{40}$$

$$= 0.025 \ \mu F$$

This capacitance is in parallel with C_c, so the combination has a value of

$$C_{abc} = C_{ab} + C_c$$
$$= 0.025 + 0.0015$$
$$= 0.0265 \ \mu F$$

This effective capacitance is in series with capacitor C_d, so the total effective capacitance for the network as a whole is

$$\frac{1}{C_t} = \frac{1}{C_{abc}} + \frac{1}{C_d}$$

$$= \frac{1}{0.0265} + \frac{1}{0.22}$$

$$= 37.74 + 4.54$$

$$\frac{1}{C_t} = 42.28$$

$$C_t = \frac{1}{42.8}$$

$$= 0.024 \ \mu F$$

RC time constants

Often a resistor and a capacitor will be combined, especially in a timing circuit. If a resistor and a capacitor are connected in series across a voltage source, as illustrated in Fig. 3-15, the capacitor will be charged through the resistor at a specific, predictable rate.

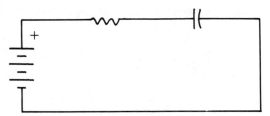

Fig. 3-15   *RC networks are often used to set up time constants.*

This charging rate is determined by both the capacitance and resistance values in the RC network. The capacitance, of course, controls how many electrons the negative plate can hold when it is fully charged, and the resistance "slows down" the flow of electrons.

The time it takes for the capacitor to be charged to 63% of its full potential is called the RC time constant. For rather obvious reasons, such a combination of a resistor and a capacitor is known as an RC circuit, or an RC network. These two terms can be considered interchangeable.

The 63% charge level was not arbitrarily selected. This particular point happens to make the mathematics involved nice and neat. The RC time constant—that is, the time it takes the capacitor to charge through the resistor to 63% of its full capacity—is equal to the resistance multiplied by the capacitance:

$$T = RC$$

where

    T = time constant, in seconds;
    R = resistance, in ohms; and
    C = capacitance, in farads.

The same formula will also work if the resistance (R) is given in megohms and the capacitance (C) is given in microfarads. But be very careful not to get confused. Do not attempt to combine ohms and microfarads, or megohms and farads. You will not get the correct results if you mix values.

As an example of how to use this formula, let's assume we are using a 100-kΩ (100,000-Ω) resistor in series with a 10-μF (0.00001-F) capacitor. In this case, the time constant of the RC network would be equal to

$$T = RC$$
$$= 100,000 \times 0.00001$$
$$= 1 \text{ second}$$

You should realize that the same time constant could be achieved with other RC combinations. For instance, a 1-MΩ (1,000,000-Ω) resistor and a 1-μF (0.000001-F) capacitor would also combine for a time constant of 1 second. So would a 5-kΩ (5,000-Ω) resistor and a 200-μF (0.0002-F) capacitor.

As another example, let's try using a 470-kΩ (470,000-Ω) resistor and a 2.2-μF (0.0000022-F) capacitor. The time constant in this circuit would be equal to

$$T = 470,000 \times 0.0000022$$
$$= 1.034 \text{ seconds}$$

For our next example, we will use a 3.3-MΩ (3,300,000-Ω) resistor and a 47-μF (0.000047-F) capacitor to give a time constant of

$$T = 3,300,000 \times 0.000047$$
$$= 155.1 \text{ seconds}$$
$$= 2 \text{ minutes, 35.1 seconds}$$

Now, here's one last example. This time the resistor has a value of 22 kΩ (22,000 Ω) and the capacitor has a value of 0.33 μF (0.00000033 F). These component values combine to give a time constant of

$$T = 22,000 \times 0.00000033$$
$$= 0.00726 \text{ second}$$
$$= 7.26 \text{ ms}$$

As you can see, a simple RC network can be set up to cover a very wide range of time constants.

In practical circuit design, you will often know the desired time constant and will have to find suitable component values. This is easy enough. Just arbitrarily select a likely capacitance value and rearrange the time constant formula to solve for the unknown resistance:

$$R = \frac{T}{C}$$

As an example, let's say we need a time constant of 2.5 seconds. If we use a 10-μF (0.00001-F) capacitor, the necessary resistance will be

$$R = \frac{2.5}{0.00001}$$
$$= 250,000 \ \Omega$$
$$= 250 \ k\Omega$$

Depending on the required precision, you could round this off to a standard 270-kΩ resistor, or you could use a 220-kΩ resistor in series with a 33-kΩ resistor (giving a total value of 253,000 Ω for R). If the application is critical, the best accuracy can be obtained by using a trimpot. This allows you to adjust the resistance to give a precise desired time constant. In most practical timing applications, fairly tight component tolerances are highly desirable, if not essential, for both the resistor and the capacitor in an RC timing circuit.

So far we have only been considering the time constant as the time it takes for the capacitor to charge (to 63%) through the resistor. The situation is very similar for discharging the capacitor. Once the capacitor is fully charged, removing the voltage source from the circuit, as illustrated in Fig. 3-16, will permit the capacitor to be discharged through the resistor. The exact same time constant is at work here. In this case, the time constant is the time it takes the capacitor to discharge through the resistor down to 37% of its fully charged level. The charging time constant and the discharging time constant are always exactly equal for any specific combination of a resistor and a capacitor.

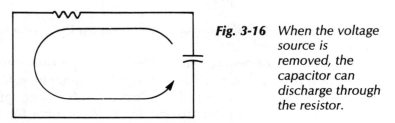

Fig. 3-16 *When the voltage source is removed, the capacitor can discharge through the resistor.*

Capacitor markings

Value markings on capacitors are not nearly as standardized as those on resistors. Several different marking systems are widely used. Many capacitors, especially those used in some commercially manufactured electronic equipment, may not be marked at all.

Clearly, a good digital capacitance meter can be a big help to most electronics hobbyists and technicians. An unknown capacitance can be quickly and easily found, without much fuss or bother. Just hook up the unknown capacitor to the instrument's test leads and read the measured capacitance value directly from the front panel of the capacitance meter.

On many capacitors, especially those with larger body sizes, such as electrolytic capacitors and relatively large ceramic disc capacitors (capacitor types will be discussed later in this chapter), the actual value may be stamped or printed directly on the body of the component. For example, an electrolytic capacitor may be marked "50 V, 470 μF." This is perfectly straightforward. The capacitor in question has a nominal capacitance of 470 μF, and the maximum voltage that can be safely applied across this device is 50 V.

On some capacitors the voltage rating may be specified as so many WV. This abbreviation represents working voltage and amounts to pretty much the same thing as the straight V rating.

A few capacitors may also indicate that the marked working voltage is a dc voltage. This is somewhat unnecessary because the voltage limit rating for any capacitor is assumed to be a dc voltage unless otherwise indicated. This means that you must be careful when using a capacitor in a circuit carrying ac signals. You must consider the peak voltage, not the rms (root mean square) voltage. For a sine wave, these two voltages interrelate as follows:

$$E_{peak} = 1.41 \times E_{rms}$$
$$E_{rms} = 0.707 \times E_{peak}$$

As an example, let's say we have a capacitor rated for a working voltage of 50 V. Can we safely use this component in a circuit carrying 37 V rms? We must convert the rms value into the appropriate peak voltage:

$$E_{peak} = 1.41 \times 37$$
$$= 52.17 \text{ V}$$

No, we'd need a heavier capacitor in this particular circuit.

Sometimes you may encounter another voltage rating for capacitors. This is usually called the breakdown voltage. If the breakdown voltage is exceeded across the plates of a capacitor, the dielectric will be punctured, and the capacitor will be totally ruined. The breakdown voltage is usually much, much higher than the working voltage rating. Most capacitors have breakdown voltages well into the kilovolt (thousands of volts) range.

Some capacitors are simply too small to permit the full value to be legibly printed on them directly. To fit more information into the available space, systems of abbreviations are often used. In one abbreviation system, the first two significant digits of the capacitance value are combined with the letter R, which indicates the position of the decimal point. For example, a capacitor might be marked R22, representing 0.22, or 3R3, indicating 3.3.

But 0.22 or 3.3 what? Is the capacitance measured in microfarads or picofarads? This is not directly indicated by the capacitor's markings. Fortunately, this isn't too much of a problem in practical electronics work. It doesn't take long for an electronics technician or hobbyist to learn to tell at a glance whether it is a 3.3-μF capacitor or a 3.3-pF capacitor, simply from the physical size of the component. Of course, the capacitor measured in microfarads will be considerably larger than a capacitor measured in picofarads.

Some capacitor manufacturers use a three-digit numerical code that is somewhat similar in concept to the standard resistor color code described in the preceding chapter. The first two digits in this code are the two most significant digits of the capacitance value. The third digit is a multiplier value. The multiplier indicates the number of zeros to follow the significant digits. For example, a code of 473 would indicate a capacitance value of 47,000. Once again, the physical size of the capacitor in question is the only direct indication of whether the indicated capacitance value is measured in microfarads or picofarads. This is usually perfectly sufficient, however, at least for experienced electronics technicians and hobbyists.

As it happens, this particular coding system will most likely be used on picofarad capacitors. The third digit (multiplier) in the code is only useful if the capacitance value is at least 100 (the code would read 101). A capacitor of 100 μF or larger would probably have enough space on its body for a printout of the full value, without resorting to any abbreviation code.

Sometimes capacitors are marked with a color code similar to the system used with resistors (described in chapter 3). This capacitor color-coding system isn't too widely used today, but you may still run across it occasionally, especially if you do much work with older electronic equipment.

This color code was used mainly with mica and paper capacitors. (Capacitor types will be discussed a little later in this chapter.) In this coding system, the body of the capacitor is marked with six colored dots arranged in two rows of three each, as shown in Fig. 3-17. The capacitor color code for each dot position is summarized in Table 3-2.

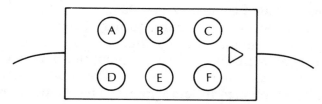

Fig. 3-17 *Some mica and paper capacitors are marked with a color-code system.*

Table 3–2   Capacitor color codes.

Color	Dot A	Dot B	Dot C	Dot D	Dot E	Dot F
Black	Mica	0	0	±1000	±20%	×1
Brown	—	1	1	±500	±1%	×10
Red	—	2	2	±200	±2%	×100
Orange	—	3	3	±100	±3%	×1000
Yellow	—	4	4	−20 to +100	—	×10,000
Green	—	5	5	0 to +70	±5%	—
Blue	—	6	6	—	—	—
Violet	—	7	7	—	—	—
Gray	—	8	8	—	—	—
White	Mica	9	9	—	—	—
Gold	—	—	—	—	±5%	×0.1
Silver	Paper	—	—	—	±10%	×0.01

The first dot, in the upper left-hand corner, is dot A. This dot is usually either white or black. This simply indicates that this component is a mica capacitor. If dot A is silver, the device in question is a paper capacitor.

Dot B is the first significant digit for the capacitance value, and dot C is the second significant digit. Notice that the color values for dot B and dot C are identical to the significant digit color values of the more widely employed resistor color code.

The multiplier for the capacitance is indicated by dot F, directly below dot C in the lower right-hand corner—not dot D at the beginning of the second row, as might be expected. Dot F corresponds to the third (multiplier) band on a resistor. The color values used for the multiplier in the capacitor color code are basically the same as in the resistor color code, but capacitors generally don't have as wide a value range as resistors, so some colors aren't used as multipliers in this system. Multiplying the two significant digits (dot B and dot C) by the multiplier (dot F) gives the capacitance value in picofarads. If you would prefer to express the capacitance value in microfarads instead of picofarads, divide the indicated value by 1,000,000.

Dot D, in the lower left-hand corner of the capacitor, gives the component's temperature coefficient. The capacitance will vary somewhat with changes in temperature. The temperature coefficient tells you the maximum extent of capacitance fluctuation for a given change in temperature. The temperature coefficient rating is given in parts per million per degree centigrade (ppm/°C). For example, if dot D is red, then the temperature coefficient of the capacitor in question is ± 200. This means that if the temperature changes 1°C, the actual capacitance value may increase or decrease by no more than 200 ppm, or 0.02%. If the temperature changes 10°C, the change in capacitance can be as high as 2,000 ppm, or 0.2%.

Notice that some capacitors have a nonsymmetrical temperature coefficient. They can increase in value more than they can decrease in value with changes in temperature. In fact, if dot D is green, the capacitance value can only increase. Changing the temperature will not cause this capacitor's value to drop. In this particular case, the capacitance value marked is said to be the minimum guaranteed value.

Dot E gives the manufacturing tolerance of the capacitor. That is, at room temperature, the actual capacitance of the component may be off from the nominal value by no more than the indicated tolerance percentage. Dot E, of course, corresponds to the fourth (tolerance) band in the resistor color code.

On some other capacitors, which do not employ the color code, a simple letter code may be used to indicate the component's tolerance rating. The tolerance letter codes are summarized in Table 3-3.

The EIA (Electronics Industries Association) has developed a letter-number-letter code to indicate the tolerance and temperature coefficient of capacitors. This code is summarized in Table 3-4. As an example of how this code works, let's say we have a

Table 3-3   Capacitor tolerance codes.

B	+0.1 pF/−0.1 pF
C	+0.25 pF/−0.25 pF
D	+0.5 pF/−0.5 pF
F	+1%/−1%
G	+2%/−2%
J	+5%/−5%
K	+10%/−10%
M	+20%/−20%
Z	+80%/−20%

Table 3-4   EIA Class II capacitor codes.

First letter code	Low temperature limit
Z	+10°C
Y	−30°C
X	−55°C
Number symbol code	**High temperature limit**
2	+45°C
4	+65°C
5	+85°C
6	+105°C
7	+125°C
Second letter code	**Maximum capacitance change**
A	±1.0%
B	±1.5%
C	±2.2%
D	±3.3%
E	±4.7%
F	±7.5%
P	±10.0%
R	±15.0%
S	±22.0%
T	+22%/−33%
U	+22%/−56%
V	+22%/−82%

capacitor that is marked Y5P. This particular code indicates that the capacitance will vary no more than 10% as long as its temperature is maintained between −30° and +45°C. Outside the specified temperature range, the actual capacitance value might change by more than the indicated tolerance percentage.

Many capacitors are marked NP0. This is often read as "En-Pe-Oh," even though the third letter is actually the numeral zero. NP0 means negative-positive-zero, indicating that the positive and negative temperature coefficients of this particular capacitor are both zero. In other words, a capacitor marked NP0 does not change its capacitance value with normal fluctuations in temperature. This can be quite important in many frequency-sensitive applications.

Classification of capacitors

Earlier in this chapter we saw that one of the major factors that determines the capacitance of a capacitor is the material used for the dielectric. A number of different materials are commonly used in capacitor dielectrics. Besides affecting the capacitance, the dielectric material also contributes to other characteristics of a capacitor. For example, some dielectric materials offer less leakage than others.

Capacitors are generally classified according to their dielectric material, although there are some exceptions. Some of the more common types of capacitors will be discussed on the following pages.

Ceramic capacitors

Perhaps the most common type of capacitor in all of electronics consists of a wafer of ceramic material between two silver plates. Leads are connected to the plates, and the entire assembly (except for the leads, of course) is encased in a protective plastic housing, which is usually (though not always) a round, flat disc, as illustrated in Fig. 3-18. This type of component is called a

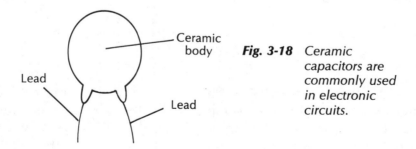

Fig. 3-18 *Ceramic capacitors are commonly used in electronic circuits.*

ceramic capacitor. Because of its shape, the ceramic capacitor is also commonly known as a ceramic disc capacitor, or simply a ceramic disc. Occasionally, a ceramic disc capacitor might be identified as a disc capacitor. This is just an alternate name for the same component. A cross-section of a typical ceramic disc capacitor is shown in Fig. 3-19.

While the term ceramic disc capacitor is commonly used in electronics for almost any ceramic capacitor, some ceramic capacitors are actually enclosed in rectangular, rather than round, cases. The difference is purely cosmetic, sometimes relating to manufacturing processes. There is no electrical difference between a true ceramic disc capacitor and a rectangular ceramic capacitor. Such rectangular ceramic capacitors are often called ceramic discs anyway.

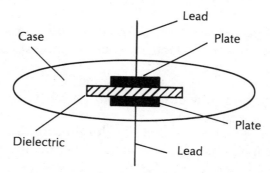

Fig. 3-19 *This is a cross-section of a typical ceramic disc capacitor.*

One of the primary reasons for the popularity of ceramic capacitors is that they are very inexpensive to manufacture. This price difference extends to the retail level and they are usually sold for considerably less than other types of capacitors.

The range of available capacitance values for ceramic capacitors is pretty wide, extending from about 10 pF to about 0.1 μF. Occasionally you might find ceramic capacitors with slightly larger or smaller capacitance values. I've seen ceramic capacitors as small as 5 pF and as large as 0.5 μF and 0.68 μF. These extreme values are very uncommon.

The working voltage rating for ceramic capacitors is usually fairly large, especially considering the low voltages found in most modern solid-state circuits. Ceramic capacitors typically have working voltages ranging from 50 V to about 1,600 V (1.6 kV). If a lower voltage unit can do the job, it will generally be less expensive, but a ceramic capacitor with a higher working voltage can always be substituted.

If you are working on an electronic project that calls for a ceramic capacitor of a specific working voltage, or if you are replacing a ceramic capacitor in an existing piece of equipment, consider the listed working voltage rating as the minimum acceptable value for the new capacitor. For example, if the project plans call for a ceramic capacitor with a working voltage of 100 V, you can safely use a 150-V capacitor, a 200-V capacitor, or even a 1,000-V capacitor, assuming that the selected device has the correct capacitance value. However, it would be a mistake to substitute a capacitor with a working voltage of only 50 V or 75 V.

As a rule, the higher the working voltage or the capacitance value, the larger the physical size of the capacitor. The size difference is generally most noticeable with an increase in the working voltage rating.

Large-value ceramic capacitors (above about 0.1 μF) usually have lower working voltage ratings. The working voltage for these relatively large capacitance units is generally only about 10 to 100 V. This is primarily because the physical size of the component would get too unwieldy if it had both a large capacitance and a large working voltage.

The capacitance value, and usually the working voltage, are normally stamped directly onto the body of ceramic capacitors. Sometimes the tolerance and temperature coefficient ratings may also be included. Sometimes one of the abbreviation codes described earlier in this chapter may be used, but it is more common for the capacitance value to given directly, especially on moderate- and large-capacitance units.

The ceramic capacitor has several important advantages. It is quite inexpensive, and it is usually fairly small (compared to most other capacitor types). Ceramic capacitors also offer relatively high reliability and a fairly low degree of power loss (leakage resistance).

However, ceramic capacitors also have numerous disadvantages. They tend to have rather wide tolerances, especially when compared with the more recently developed synthetic-film capacitors (described later in this chapter). The stability of ceramic capacitors tends to be rather poor, especially at higher frequencies.

On some ceramic capacitors a simple letter code is used to indicate the component's tolerance. There are four standard code letters used:

- M = ± 20%;
- K = ± 10%;

- J = ± 5%; and
- C = ± 0.25 pF.

If the capacitor's tolerance code is C, the actual capacitance won't deviate from the nominal value by more than 0.25 pF. This tolerance code is found only on very small-value capacitors with nominal capacitances of just a few picofarads.

As an example, we will consider the maximum allowed capacitance deviation for each of these tolerance codes. We will assume that the capacitor's nominal value in each case is 22 pF. If the tolerance code is M, the actual capacitance can be as low as 17.6 pF or as high as 26.4 pF. A tolerance code of K would indicate that the actual capacitance would be no lower than 19.8 pF and no higher than 24.2 pF. If the tolerance code is J, the actual capacitance will be between 20.9 and 23.1 pF. Finally, if the capacitor's tolerance code is C, the actual capacitance value can be no lower than 21.75 pF and no higher than 22.25 pF. Of course, a tighter tolerance rating usually means a more expensive capacitor.

The ceramic capacitor's poor high-frequency stability makes it an unsuitable choice for rf circuits. A ceramic capacitor can usually be expected to do a pretty good job at frequencies below about 100 kHz (100,000 Hz), but operation will become erratic if the signal frequency is increased beyond this point.

Different component manufacturers may use slightly different capacitance values. Because of the relatively wide tolerances of most ceramic capacitors, it doesn't make much sense to knock yourself out looking for an exact value. Feel free to round off the desired capacitance to the nearest available value. For instance, some manufacturer's might make 0.0047-μF ceramic capacitors, while another manufacturer's comparable components may be marked 0.005 μF. The difference in values between these two capacitors is negligible. Similarly, a 0.02-μF capacitor, a 0.022-μF capacitor, and a 0.25-μF capacitor could probably all be considered to be pretty much interchangeable in a practical electronics circuit. If high precision in the capacitance value is required, a ceramic capacitor probably isn't the best choice anyway.

There are several variations on the basic ceramic disc capacitor. For instance, ceramic chip capacitors have become quite popular among some manufacturers of electronic equipment. This type of component is also sometimes called a printed circuit capacitor. Unlike most other capacitors (and other electronic components), ceramic chip capacitors often have no leads. They are designed specifically for surface mounting on a printed circuit board. They are difficult to install by hand, but in a factory assembly line they are very convenient.

The chief advantage of the ceramic chip capacitor is its very small physical size, coupled with a moderately large capacitance. The relatively large capacitance within such a small space is accomplished by sandwiching multiple layers of the ceramic dielectric and alternating plates. The multiple plates are connected in parallel, forming two fairly large plates (in terms of plate-to-dielectric contact area).

The specific ceramic material used as the dielectric in ceramic chip capacitors has a very, very high dielectric constant—typically between 2000 and 6000. This extremely high dielectric constant also helps boost the capacitance.

Another important space-reduction technique used in the manufacture of ceramic chip capacitors is to replace the foil sheets normally used as capacitor plates with a metallic ink printed directly onto the surface of the ceramic dielectric. This makes the sandwich significantly thinner overall.

Mica capacitors

Another popular type of capacitor is the mica capacitor, which features thin slices of mica (the dielectric) sandwiched between a number of interconnected plates. The basic internal construction of a mica capacitor is illustrated in Fig. 3-20. Again, this type of capacitor is named after its dielectric material.

As a rule, mica capacitors are somewhat more expensive than ceramic disc capacitors, but they also tend to be much more stable at radio frequencies. Mica capacitors also have fairly good capacitance-value tolerances. Tolerance ratings for mica capacitors are usually 10% or 20%. Precision mica capacitors with even tighter tolerances are also available.

The available capacitance values for mica capacitors range from 5 pF to about 0.01 μF. Working voltages for mica capacitors generally run between 200 V and 50,000 V (50 kV).

Mica capacitors, especially in older electronic equipment, are often marked with the color-code system described earlier in this chapter. Other mica capacitors have their capacitance value stamped directly onto the body of the device, usually with one of the abbreviation codes introduced earlier in this chapter.

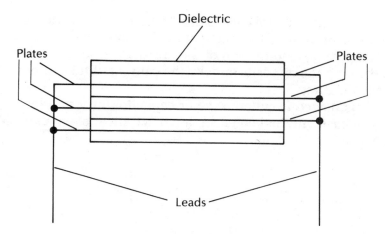

Fig. 3-20 *Mica capacitors are also popular in electronic circuits.*

A special type of mica capacitor is the silver-mica capacitor. This component is also sometimes called a silvered mica capacitor. In this type of capacitor a very thin coating of silver, instead of the usual aluminum foil plates, is applied directly onto the mica sheets. Silver, of course, is a better conductor than aluminum foil.

Naturally, silver-mica capacitors are more expensive than regular mica capacitors. The advantage of the silver-mica capacitor is that very tight capacitance tolerances are possible. Some silver-mica capacitors have tolerances as close as 1%. This is clearly a significant improvement over the ordinary mica capacitor or the ceramic capacitor.

The high cost of silver-mica capacitors is usually justified only in critical applications requiring precise time constants or frequencies (especially in the rf region).

Paper capacitors

Sometimes an electronic circuit will require a higher capacitance than a ceramic disc capacitor or a mica capacitor can provide. Another popular type of capacitor is formed by placing a strip of waxed paper between two strips of conductive foil (the plates). These foil and paper strips are often several feet long, but only an inch or less wide. In effect, we have a very long, flat capacitor. This gives us a fairly large capacitance, but it takes up a lot of space. To save space, this long flat capacitor is tightly rolled and enclosed in a protective cardboard or plastic tube. Sometimes a

sealed plastic or metal case is used to house the capacitor to prevent moisture from getting inside the component. Of course,
leads are connected to each of the foil strips (plates) and brought
out through the opposite ends of the cardboard tube or other protective housing.

This device is called a paper capacitor, or a tubular capacitor. The basic internal construction of a typical paper capacitor is
shown in Fig. 3-21.

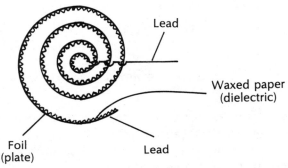

Fig. 3-21   *A paper capacitor can offer a moderately large capacitance in a
relatively small space.*

Because of the large plate-to-dielectric contact area, paper
capacitors can offer fairly large capacitances. The capacitance
range for most paper capacitors runs from about 0.0001 μF to
approximately 1 μF. Typical working voltages for paper capacitors are generally in the 200 V to 5,000 V (5 kV) range.

Paper capacitors are usually relatively inexpensive, but they
have a fairly low leakage resistance. A paper capacitor has a
much lower leakage resistance rating than either a mica capacitor
or a ceramic capacitor. In some applications, this might not matter, but in other applications, the paper capacitor might be an
unsuitable choice.

A paper capacitor does not give optimum performance at the
upper radio frequencies because of high dielectric losses. However, paper capacitors are generally an excellent choice when a
fairly large amount of capacitance is needed in a low- to medium-
frequency circuit, especially when space is at a premium.

Paper capacitors are generally large enough to have their values stamped or printed directly onto the body of the component.
Some paper capacitors (especially in older electronic equipment)

are marked with the color-coding system more commonly used with mica capacitors. (This color code was discussed earlier in this chapter.) When a paper capacitor uses this color code, dot A (in the upper left-hand corner) is always silver, indicating the capacitor type.

Synthetic-film capacitors

A capacitor similar to the paper capacitor can be made by using a thin layer of synthetic film or plastic as the dielectric instead of waxed paper. Such synthetic-film capacitors are a relatively recent development.

A capacitor with a synthetic-film dielectric is usually identified by the specific type of plastic film employed. Some typical synthetic-film capacitors include polystyrene capacitors, Mylar capacitors, polyester capacitors, and polypropylene capacitors, among others. These synthetic films are usually significantly thinner than the waxed paper used in paper capacitors, so a synthetic-film capacitor of a given capacitance value will tend to be somewhat smaller physically than a comparable paper capacitor.

Another important advantage of these synthetic-film capacitors is that they can generally operate reliably over a wider temperature range than paper capacitors can handle. In other words, synthetic-film capacitors have a lower temperature coefficient than paper capacitors. Synthetic-film capacitors also tend to have very tight tolerances. The actual capacitance value is usually very close to the nominal value printed on the body of the capacitor.

Because synthetic-film capacitors tend to be fairly small, the abbreviation codes described earlier in this chapter are quite frequently used to mark the capacitance value on this type of capacitor. Synthetic-film capacitors are available with capacitances ranging from 0.001 μF to about 2 μF. The working voltage rating for this type of capacitor is generally between 50 V and 1000 V.

Don't be thrown by the many different types of synthetic-film capacitors that are now available. While there are some minor functional differences in capacitors using different types of plastic film as their dielectrics, in most practical electronic applications, these differences are quite negligible. With very few exceptions, all synthetic-film capacitors can be considered more or less interchangeable.

I remember one time, when I was just getting started as an electronics hobbyist, I was working on a project from one of the electronics hobby magazines. The parts list called for a couple of polystyrene capacitors. I spent weeks trying to locate these components with absolutely no luck at all. Eventually, I ended up substituting Mylar capacitors for the polystyrene capacitors called for in the parts list. At the time I thought I was taking a chance, but this kind of substitution is practically irrelevant. The circuit didn't care. The project worked just fine with the Mylar capacitors. If I had been able to locate the specified polystyrene capacitors, it is doubtful they would have made any noticeable difference in the operation of the circuit.

On the other hand, substituting a ceramic disc capacitor or a paper capacitor for a polystyrene (or other synthetic-film) capacitor may or may not make a difference. In this case you'd be gambling on the component's tolerance, stability, and temperature coefficient. You can always substitute a more precise (lower tolerance) component in a less critical application, but making the opposite substitution can be risky.

Of course, you can use a synthetic-film capacitor in place of a ceramic disc capacitor, but this is rarely very practical. Generally a synthetic-film capacitor will cost a lot more than a comparable ceramic capacitor, and it isn't likely that the substitution will make any noticeable improvement in the operation of the circuit.

Electrolytic capacitors

Ceramic disc capacitors, mica capacitors, paper capacitors, and synthetic-film capacitors are all nonpolarized devices. This means they can be hooked up in the circuit without regard to polarity. These capacitors don't care which plate is positive and which is negative.

Other capacitors have a definite polarity. They are designed to work only in dc circuits and can be hooked up only one way. Such a capacitor is said to be polarized.

The symbol for a polarized capacitor is basically the same as for a regular capacitor, but a plus sign (+) is added to the positive terminal to indicate the polarity of the device. The flat side of the capacitor symbol is always used as the positive terminal. This terminal must be kept at a higher electrical potential (more positive voltage) than the other terminal. The polarized capacitor symbol is shown in Fig. 3-22. The positive lead will also be marked on the body of the polarized capacitor, permitting the technician or

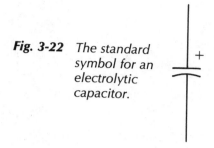

Fig. 3-22 *The standard symbol for an electrolytic capacitor.*

electronics hobbyist to correctly determine the component's polarity.

The most common type of polarized capacitor is the electrolytic capacitor. This type of capacitor consists of pasty, semiliquid electrolyte between aluminum foil electrodes or plates. Internally, an electrolytic capacitor resembles the construction of a paper capacitor, except just one electrode is used and the dielectric is a thin oxide film. One side of the electrode (the positive plate) is specially treated to form this thin oxide film on its surface. If the capacitor is subjected to a reverse polarity voltage, this thin layer of film could be punctured, ruining the capacitor.

Because the dielectric is such a thin film and only one electrode is used, very high capacitance values can be achieved in a reasonably small space. Even so, some electrolytic capacitors are quite large when compared to other types of capacitors.

Typical values for electrolytic capacitors range from 0.47 μF to 10,000 μF (0.01 F), and occasionally even larger. Working voltage ratings for this type of component can be anywhere from 3 V to 700 V. Because of their relatively large physical size, electrolytic capacitors almost always have their capacitance value and working voltage stamped directly onto the body of the component.

Electrolytic capacitors are usually enclosed in sealed metallic (or sometimes plastic) cans to protect the semiliquid electrolyte. The leads on an electrolytic capacitor might be placed in an axial arrangement, as shown in Fig. 3-23, or in a radial configuration, as illustrated in Fig. 3-24. In either case a small plus sign (+) is printed on the body of the capacitor to indicate which is the positive lead.

Electrolytic capacitors almost always have extremely wide tolerances. A typical tolerance rating for an electrolytic capacitor is + 80%/ – 20%. That is, the actual capacitance value may be up

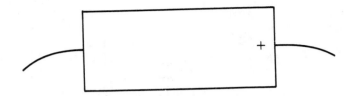

Fig. 3-23   *Some electrolytic capacitors are packaged with axial leads.*

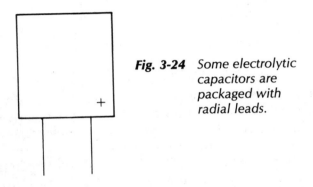

Fig. 3-24   *Some electrolytic capacitors are packaged with radial leads.*

to 80% higher or 20% lower than the nominal value printed on the body of the capacitor. For example, let's consider a 220-μF electrolytic capacitor. The actual capacitance of this component may be as low as 176 μF or as high as 396 μF. Obviously an electrolytic capacitor is not suitable for applications requiring a very high degree of precision.

If a reverse polarity voltage is applied across an electrolytic capacitor it could be ruined or the capacitor might even explode. It might seem that the electrolytic capacitor would be of rather limited value because of its inability to function in an ac circuit. Actually, this type of component has a number of important applications. These include power-supply filtering, circuit coupling, audio-frequency bypassing, and achieving large time constants in RC circuits. Of course, in a timing circuit requiring any precision, a variable resistance (a potentiometer) should be used to compensate for the very wide tolerance of the electrolytic capacitor.

One problem with electrolytic capacitors is that the semiliquid electrolyte can dry out, rendering the capacitor useless. This generally occurs when the capacitor is not in use for an extended period of time. Applying a voltage across an electrolytic capacitor seems to prevent the electrolyte from drying out. For most electronic components, it is economical to buy relatively large quantities and stockpile them for later use. This is not a good idea for electrolytic capacitors because they can go bad while waiting on the shelf.

Similarly, drying out of the electrolyte can occur if the electrolytic capacitor is operated at too low a voltage. If the operating voltage is extremely low (in comparison with the component's working voltage rating), then as far as the capacitor is concerned, it is still sitting on the shelf. Certainly a 15-V electrolytic capacitor can be used in a 5-V circuit, but you probably wouldn't want to use a 500-V unit. Besides, electrolytic capacitors with higher working voltages are much larger, heavier, and more expensive than smaller units.

Preferably, an electrolytic capacitor should be operated at a voltage between one-third and two-thirds of its maximum working voltage rating. This is enough voltage to keep the electrolyte from drying out, but allows some headroom for unexpected overvoltage surges.

The housing of an electrolytic capacitor is normally sealed and airtight. This helps minimize electrolyte problems due to environmental conditions. Sometimes the seal might be broken in some way. This will shorten considerably the anticipated lifespan of the component.

Occasionally, when working on older electronic equipment, you may find a whitish powder around or caked to the body of an electrolytic capacitor. This powder is the dried-out electrolyte that has somehow leaked out.

Nonpolarized electrolytics

In some practical electronics applications, we might need the high-capacitance values of an electrolytic capacitor in an ac circuit. But an electrolytic capacitor is polarity sensitive. A nonpolarized electrolytic capacitor can be created by placing two electrolytic capacitors in parallel, as illustrated in Fig. 3-25. To minimize the possibility of problems, each of these electrolytic

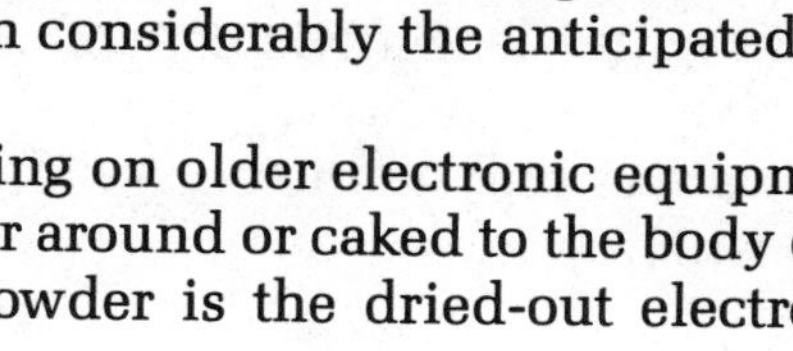

Fig. 3-25 A nonpolarized electrolytic capacitor can be made by using two ordinary electrolytic capacitors in parallel.

capacitors should be protected with a diode, as shown in Fig. 3-26. These diodes ensure that each electrolytic capacitor sees only the correct polarity. Some nonpolarized electrolytic capacitors are commercially available, usually in sizes ranging from about 1 μF to 10 μF. These components are most commonly used in crossover networks in audio speaker systems.

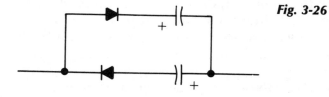

Fig. 3-26 *Better polarity protection is possible by adding a pair of diodes to the electrolytic capacitors.*

Tantalum capacitors

Closely related to the electrolytic capacitor is the tantalum capacitor. Tantalum capacitors tend to be considerably smaller than electrolytic capacitors with comparable values. The body of a tantalum capacitor is teardrop shaped, as illustrated in Fig. 3-27.

Like the electrolytic capacitor, a tantalum capacitor can be used only in dc circuits. It is a polarized component. A dot (usually white or red) on the body of the capacitor is used to indicate the positive lead.

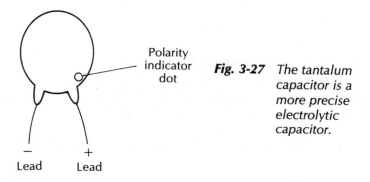

Polarity indicator dot

Fig. 3-27 *The tantalum capacitor is a more precise electrolytic capacitor.*

The basic internal construction of a typical tantalum capacitor is illustrated in Fig. 3-28. Like the electrolytic capacitor, the tantalum capacitor uses a single metallic electrode along with an electrolyte. However, in the case of the tantalum capacitor, the electrode is not a sheet of metal. Instead, a slug of tantalum powder is used as the electrode. This powder is very porous, and the

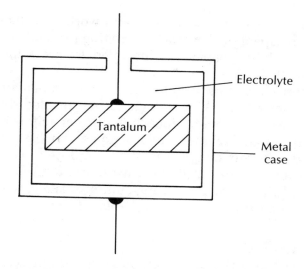

Fig. 3-28   *The basic internal structure of a typical tantalum capacitor.*

surfaces of the grains are anodized, producing tantalum oxide. Tantalum is a conductor, but tantalum oxide is an insulator, and it serves here as the capacitor's dielectric.

The slug of tantalum powder is also covered with an electrolyte in liquid form. The electrolyte solution in a tantalum capacitor contains manganese nitrate. When this electrolyte solution is absorbed by the tantalum powder and then evaporates, the remaining residue is manganese dioxide, which serves as the capacitor's active electrolyte.

Tantalum capacitors offer a number of important advantages over their electrolytic counterparts. They tend to be considerably smaller for a given capacitance value. They aren't as prone to drying out and can be stockpiled in storage. Tantalum capacitors have considerably less leakage and better long-term stability than electrolytic capacitors. Their values tend to be much more precise (lower tolerance), and they are much less susceptible to noise.

The tighter tolerances of tantalum capacitors make them suitable for more precision applications than the wider tolerances of electrolytic capacitors. Because of their rejection of noise, tantalum capacitors are generally preferred in computer applications.

On the other hand, tantalum capacitors tend to be more expensive than electrolytic capacitors, and they are limited to a much smaller range of capacitances and working voltages. Usually tantalum capacitors have values ranging from 0.5 μF to about 50 μF. Few tantalum capacitors are rated for working voltages higher than 50 V.

Tantalum capacitors usually have their capacitance value (and sometimes the working voltage rating) printed directly on the body of the component. Since tantalum capacitors tend to be rather small, an abbreviation code of some sort (like those discussed earlier in this chapter) is often employed. The units of capacitance are usually omitted to save space, but this should be no problem. If it is a tantalum capacitor, it is safe to assume that the capacitance value is given in microfarads.

Variable capacitors

By far the vast majority of capacitors have a single fixed capacitance value. But, just as there are variable resistors (see chapter 2), there are also variable capacitors.

One type of variable capacitor uses a springy material for the plates. The assembly is held together by a small screw. The plates are so springy that if they weren't held in place by the screw, they would fly apart. Between the plates is a piece of dielectric material. Mica is commonly used in this type of variable capacitor. The basic structure of this component is illustrated in Fig. 3-29. By tightening or loosening the screw holding the capacitor together, the distance between the plates is changed, thus altering the effective capacitance of the device. A variable capacitor of this type is called a padder capacitor or a trimmer capacitor.

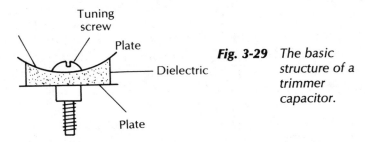

Fig. 3-29 *The basic structure of a trimmer capacitor.*

Another type of variable capacitor consists of a series of interleaved metal plates. One set of plates (called the stator) is stationary. The other set of plates (called the rotor) can be moved via a knob attached to the end of a control shaft. The amount of overlap between the two sets of plates determines their effective area, and thus the effective capacitance of the device. In other words, moving the rotor creates the electrical effect of increasing or decreasing the size of the capacitor's plates. Figure 3-30 illustrates the way this type of variable capacitor works.

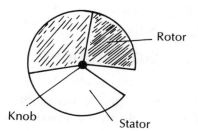

Fig. 3-30   *A variable capacitance can also be obtained from a variable air capacitor.*

The dielectric in this particular type of capacitor is air. For this reason, the component is often called a variable air capacitor or an air variable capacitor.

The symbol for any type of variable capacitor is shown in Fig. 3-31. It is just the standard symbol for a capacitor with an arrow through it, indicating a variable component.

Variable capacitors are not commonly used in modern electronics, but most technicians and electronics hobbyists do run across them from time to time.

Fig. 3-31   *The symbol for a variable capacitor.*

Problems with capacitors

Overall, capacitors are pretty reliable components, but problems can occur. Outside the realm of the capacitor's temperature coefficient, the capacitance value is normally quite constant. If you measure the capacitance of a specific capacitor periodically, you should always get the same result, unless extreme fluctuations of temperature are involved.

When a capacitor does go bad, it usually is either shorted or open. In a shorted capacitor, there is a direct dc path between the plates. The capacitor acts like a simple piece of wire, or perhaps a resistor. An open capacitor, however, has no connection at all (either ac or dc) between its plates. It acts like an open circuit, or like a very, very large resistor or capacitor.

Sometimes capacitor problems can be intermittent. That is, sometimes it will work OK, but at other times it will be either open or shorted. This is usually due to a small crack in the capacitor. Such cracks are most common with ceramic capacitors.

Stray capacitances

Since a capacitor is simply two conducting surfaces separated by an insulator, small, unintentional capacitances can often be formed in a circuit between adjacent wires or component leads. Generally, these stray capacitances are far too small to be of any real significance, but in some very high-frequency (VHF) circuits (such as radio circuits) they can be very troublesome. These undesirable capacitances can allow signals to pass into inappropriate portions of the circuit where they can hinder proper operation.

To prevent such stray capacitances in high-frequency circuits, leads should be kept as short as possible, reducing the effective plate area in the undesired pseudocapacitor. Leads should also be shielded whenever possible if they are more than a few inches long. A shielded lead is enclosed in (but insulated from) a conductor that is connected to ground potential.

Printed circuit boards for high-frequency circuits often use guard bands or extra conductive traces between the circuit traces. These guard bands help break up potential stray capacitances. The guard bands are usually shorted to ground or left unconnected to anything in the circuit.

❖ 4
Coils and transformers

THE THIRD BASIC PASSIVE COMPONENT OF ELECTRONICS IS THE coil, or inductor. Where the resistor exhibits resistance, and the capacitor exhibits capacitance, the inductor, not surprisingly, exhibits inductance. Except for some radio circuits, inductors aren't used in modern electronic circuits nearly as often as capacitors or resistors, but this type of component is still very important.

Inductance, like capacitance, is an ac phenomenon. It is frequency sensitive. That is, signals at different frequencies respond to inductance differently. Just as a capacitor exhibits a type of ac resistance known as capacitive reactance, an inductor exhibits a different type of ac resistance, called inductive reactance.

The alternate name for an inductor is a coil. This is a descriptive name for this component. Usually, an inductor is just a length of wire wound into a spiral or coil shape. The coil may or may not be wound around a core of some sort.

The electromagnetic effect

When an electric current passes through a conductor, such as a piece of copper wire, a weak magnetic field is produced around the conductor. The magnetic lines of force encircle the wire at right angles to the direction of current flow and are evenly spaced along the length of the conductor. This is part of what is known as the electromagnetic effect.

The strength of the magnetic field decreases at greater distances from the conductor. The size and overall strength of the magnetic field is dependent on the amount of power flowing through the electrical circuit, but it is always fairly weak with a straight conductor.

The magnetic force surrounding the conductor can, however, be significantly increased by winding the wire into a coil, so the lines of magnetic force can interact and reinforce each other. An even greater magnetic force can be generated if the coil is wound around a core made of low-reluctance material, such as soft iron. Reluctance is basically the magnetic equivalent of electrical resistance.

The electromagnetic effect also works in the opposite direction. That is, not only can we produce magnetism with an electrical current, we can also produce electricity with a magnet.

If we move a magnet up through a coil of wire, an electric current will start to flow through the wire. The strength of this induced current depends on a number of factors, including the intensity of the magnetic field, how many lines of magnetic force are cut by the conductor, the number of conductors (each turn of the coil acts like a separate conductor in this case) cutting across the lines of force, the angle at which the lines of force are cut, and the speed of the relative motion between the magnet and the conductor. The induced electrical current will continue to flow until either the magnet is too far away for any of its lines of force to cut across the conductor or the magnet stops moving.

If the magnet and the coil are stationary with respect to each other, no current is induced. The relative motion is an essential part of the process. If we push the magnet back down through the coil, in the opposite direction than before, current will also flow through the coil, but it will have the opposite polarity. That is, the induced current flows in the reverse direction. The exact same effect can be achieved if the magnet is held stationary and the conductor is moved. It is the relative motion between the magnet and the coil that is important.

What is inductance?

Since an electric current through a coil of wire can create a magnetic field, and a magnetic field moving relative to a coil of wire can induce an electric current, what happens when the current flowing through a coil changes?

As long as current flows through the coil at a steady, constant level and in just one direction (dc), a nonmoving magnetic field is generated around the coil. As long as this magnetic field and the coil are stationary with respect to each other, the magnetic field will have no particular effect on the current flow through the coil. But if the current through the coil starts to drop, the magnetic force generated around the coil will also be decreased, causing the magnetic lines of force to move closer. In effect, the magnetic field is moving with respect to the coil. Naturally, some of these moving lines of magnetic force will cut through some of the turns of the coil, inducing an electric current in it. This induced current will flow in the same direction (same polarity) as the original current.

Of course, the induced current now passing through the coil will generate a magnetic field of its own, although this new magnetic field will be significantly weaker than the original one that induced the current in the coil. This means that some finite period of time is required for this back-and-forth effect to die down. Current through a coil cannot be instantly stopped or reversed in polarity. In other words, inductance tends to oppose any change in current flow.

In some ways, inductance is the opposite of capacitance. Capacitance offers very little resistance to high frequencies, but opposes low frequencies or dc (constant current). Inductance, on the other hand, passes dc with practically no resistance at all, but opposes higher ac frequencies (changing current). This opposition to high frequencies is called inductive reactance. Inductive reactance increases with increases in frequency.

Inductance is measured in henries (H), named after Joseph Henry (1797 – 1878), yet another pioneering scientist in the early days of electronics. One henry is the inductance in a circuit in which the current changes its rate of flow by 1 A/s and induces 1 V in the coil. The henry is far too large a unit for most practical electronic circuits, so the millihenry (one-thousandth of a henry) is more commonly used. This is abbreviated as mH. As an example, a 50-mH coil has an inductance of 0.05 H.

Inductive reactance

As you can see, we now have three forms of resistance—dc resistance (discussed in chapter 2), capacitive reactance (discussed in chapter 3), and inductive reactance. Inductive reactance depends on the inductance of the coil and the signal frequency. In most technical texts and formulas, inductive reactance is identified as X_L. The formula for inductive reactance is

$$X_L = 2\pi FL$$

where

    X_L = inductive reactance, in ohms;
    L = inductance, in henries;
    F = frequency, in hertz; and
    π = mathematical constant, 3.14.

Therefore 2π equals 6.28, so the inductive reactance formula can be rewritten as

$$X_L = 6.28FL$$

As an example of inductive reactance, let's suppose we have a circuit with 100 mH (0.1 H) of inductance. If the frequency of the source voltage is 60 Hz, then the inductive reactance equals

$$X_L = 6.28 \times 60 \times 0.1$$
$$= 37.68 \ \Omega$$

Now, let's suppose we have the exact same circuit, but the signal frequency is increased to 500 Hz. The inductive reactance in this case works out to a value of

$$X_L = 6.28 \times 500 \times 0.1$$
$$= 314 \ \Omega$$

Raising the signal frequency still further to 2000 Hz brings the inductive reactance to a value of

$$X_L = 6.28 \times 2000 \times 0.1$$
$$= 1256 \ \Omega$$

Incidentally, a dc voltage has a frequency of 0 Hz, so the inductive reactance of our hypothetical circuit is

$$X_L = 6.28 \times 0 \times 0.1$$
$$= 0 \ \Omega$$

This will be true for any inductance value. Of course, in any practical component there will be some dc resistance, although the value will be quite low. But the inductive reactance itself will be zero for a dc signal.

If the signal frequency remains constant, but the inductance is increased, then the inductive reactance will also be increased.

For example, we've already found that applying a 60-Hz signal to a circuit with an inductance of 100 mH results in an inductive reactance of about 38 Ω. If we increase the inductance to 500 mH (0.5 H) and keep the signal frequency at 60 Hz, the inductive reactance becomes

$$X_L = 6.28 \times 60 \times 0.5$$
$$= 188.4 \ \Omega$$

Increasing either the signal frequency or the inductance increases the inductive reactance. You should notice that the relationship of the frequency and the inductance to the inductive reactance is just the opposite of the relationship of frequency and capacitance to capacitive reactance.

When passing through an inductive reactance, the voltage leads the current by 90 degrees. Again, this is just the opposite of what happens with capacitive reactance, where the voltage lags the current by 90 degrees.

Impedance

We have three basic types of resistance, one for each of the three basic passive component types:

- Resistance (dc)—resistor;
- Capacitive reactance—capacitor; and
- Inductive reactance—inductor.

A dc voltage is only affected by the dc resistance. To a dc signal, inductive reactance looks like a dead short (zero resistance), and capacitive reactance looks like an open circuit (infinite resistance). But an ac voltage is affected by all three types of resistance in a circuit. The combined effect of all three types of resistance is the total ac resistance, or the impedance of the circuit. Naturally, since both inductive reactance and capacitive reactance are frequency dependent, so is impedance.

Impedance in inductive circuits

First we will consider impedance in a purely inductive circuit. Any practical inductor has some dc resistance, as well as its inductance and inductive reactance. It also has some capacitance (and therefore some capacitive reactance), but usually this value

is small enough to be ignored in practical electronic circuits
(unless extremely high frequencies are involved). For now, we
will ignore the inductance factor and concentrate on a purely
inductive circuit.

Let's assume we are working with the simple three-part circuit shown in Fig. 4-1. This circuit is composed of a resistance, a
coil, and an ac voltage source. The effective circuit can be

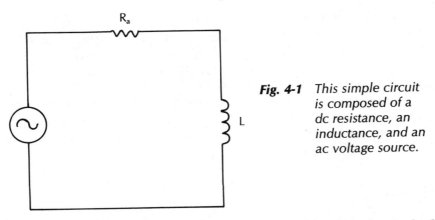

Fig. 4-1   *This simple circuit
is composed of a
dc resistance, an
inductance, and an
ac voltage source.*

redrawn as shown in Fig. 4-2. The additional resistance marked
R_L does not exist in this circuit as a separate component. It represents the internal dc resistance of the coil itself. It is shown here
to make the following discussion a little clearer.

For the following example, we will assign these values to the
components in the circuit:

- L = 100 mH (0.1 H);
- R_L = 22 Ω; and
- R_a = 1000 Ω.

We'll examine what happens in this circuit as different frequencies are applied.

If the frequency of the ac source is 60 Hz, the inductive reactance is about 38 Ω. This was one of our earlier examples. It seems
that we could find the total effective resistance of this circuit simply by adding X_L + R_L + R_a, giving a grand total of 1060 Ω. Unfortunately, this simple and "obvious" approach will not give us an
accurate or meaningful result. The reason for this is that the
inductive reactance is 90 degrees out of phase with the dc resistance.

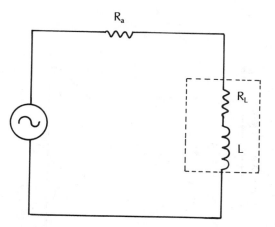

Fig. 4-2 *This is the circuit of Fig. 4-1 redrawn to show the inductor's internal dc resistance.*

An inductance offers more opposition to current than to voltage, so the voltage leads the current by 90 degrees. In a purely resistive dc circuit, the voltage and current are always in phase with each other.

Another reason why resistance and reactance cannot be added together is that reactance is not really a true resistance—it is only apparent resistance. In a true dc resistance, power is consumed by the resistor (usually the energy is converted into heat), but no power is actually consumed by a reactance (either inductive or capacitive).

We could find the impedance of the circuit (the combined effect of the resistance and the reactance) with a vector diagram, as shown in Fig. 4-3. The inductive reactance line is drawn to scale, proportional to the inductive reactance, and the resistance line is given a length proportional to the dc resistance. Dotted cross lines are drawn at right angles at the free ends of the inductive reactance and resistance lines, forming a box. A diagonal line is drawn from the point of origin to the corner of the dotted cross lines. The length of this diagonal line is proportional to the impedance of the circuit.

Since the phase relationship between resistance and inductive reactance is always 90 degrees, we can also solve the problem mathematically. The vector diagram of the resistance and the inductive reactance can always be drawn in the form of a right triangle, as shown in Fig. 4-4. The sides of the triangle have

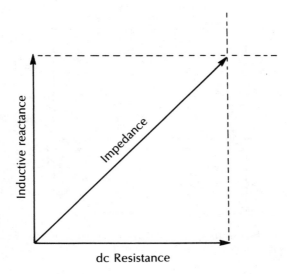

Fig. 4-3   *A vector diagram can be used to determine the impedance of a circuit with dc resistance and inductive reactance.*

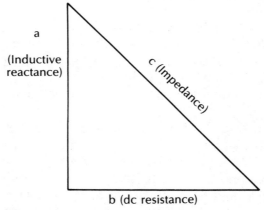

Fig. 4-4   *The vector diagram can be redrawn as a right triangle.*

the following meanings:

- a = inductive reactance;
- b = resistance; and
- c = impedance.

If we know the length of the triangle sides connected at the right angle (a and b), we can solve for the third side with the algebraic formula

$$c^2 = a^2 + b^2$$

Or we can rewrite the formula using the appropriate electrical terms:

$$Z = \sqrt{R^2 + X_L^2}$$

where

Z = impedance, in ohms;

R = resistance, in ohms; and

X_L = inductive reactance, in ohms.

Returning to our example circuit, the inductive reactance is equal to 38 Ω and the total dc resistance is equal to

$$R = R_a + R_L$$
$$= 1000 + 22$$
$$= 1022 \ \Omega$$

Therefore, the total impedance for this circuit with a signal frequency of 60 Hz is approximately

$$Z = \sqrt{1022^2 + 38^2}$$

$$= \sqrt{1,044,484 + 1444}$$

$$= \sqrt{1,045,928}$$
$$= 1022.7 \ \Omega$$

For a second example, let's keep all the same component values, but increase the signal frequency to 200 Hz. The dc resistance is not affected by the signal frequency, but the inductive reactance changes to

$$X_L = 6.28FL$$
$$= 6.28 \times 200 \times 0.1$$
$$= 125.6 \ \Omega$$

We can round this off to 126 Ω for convenience and find the new total impedance for the circuit:

$$Z = \sqrt{1022^2 + 126^2}$$

$$= \sqrt{1,044,484 + 15,876}$$

$$= \sqrt{1,060,360}$$

$$= 1030 \ \Omega \ (\text{approximately})$$

As the frequency increases, both the inductive reactance and the total impedance of the RL circuit are also increased. The value of the dc resistance remains constant, regardless of the signal frequency applied to the circuit.

Things get just a little more complicated with a parallel RL circuit like the one shown in Fig. 4-5. Here we have a resistor (R_a) in parallel with a coil. The coil's internal ac resistance is made up of the inductive reactance (X_L) in series with the component's internal resistance (R_L), as shown in Fig. 4-6.

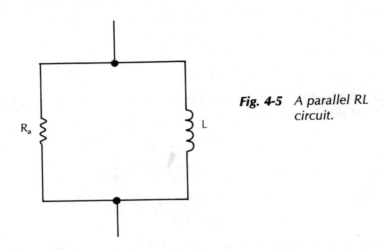

Fig. 4-5 A parallel RL circuit.

In most practical circuits, the internal resistance of the coil is so small it has only a negligible effect on the total impedance. It simplifies the equation quite a bit if we ignore R_L. The formula for the impedance of a parallel RL circuit (ignoring the effects of R_L) is

$$Z = \frac{RX_L}{(\sqrt{R^2 + X_L^2})}$$

Again, all three variables in this equation are in ohms.

As an example, let's suppose we have 500 Ω of dc resistance in parallel with an inductive reactance of 127 Ω. In this case the impedance of the RL parallel network would work out to

$$Z = \frac{(500 \times 127)}{(\sqrt{500^2 + 127^2})}$$

$$= \frac{63,500}{(\sqrt{250,000 + 16,129}\ )}$$

$$= \frac{63,500}{\sqrt{266,129}}$$

$$= \frac{63,500}{515.88}$$

$$= 123.1\ \Omega$$

Notice that the total impedance of a parallel RL circuit will always be less than either the dc resistance or the inductive reactance separately. Of course, this is similar to the usual situation of dc resistances in parallel.

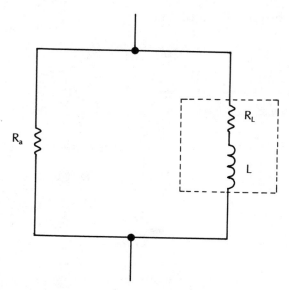

Fig. 4-6 *The circuit of Fig. 4-5 redrawn to show the internal dc resistance of the inductor.*

Separate impedances can be combined just like resistances (see chapter 2). For instance, if we have two RL combinations in series, as shown in Fig. 4-7, the impedances add:

$$Z_t = Z_1 + Z_2 + Z_3 + \ldots + Z_n$$

Fig. 4-7   *If there are two or more RL combinations in series, the impedances add.*

If we have more than one RL network in parallel, as illustrated in Fig. 4-8, the total effective impedance can be found with the formula:

$$\frac{1}{Z_t} = \frac{1}{Z_1} + \frac{1}{Z_2} + \frac{1}{Z_3} + \ldots + \frac{1}{Z_n}$$

Both of these equations assume that all of the impedances involved are in phase with each other. Dissimilar (out-of-phase) impedances cannot be combined in this manner. As long as we are dealing only with dc resistances and inductive reactances, there is no problem, but once we add in capacitive reactances, things get a bit more complicated.

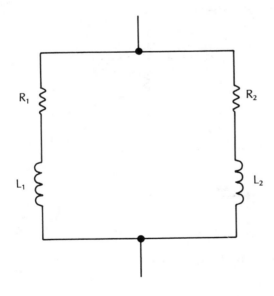

Fig. 4-8   *Series RL networks can also be combined in parallel.*

Impedances in capacitive circuits

Before we look at mixing inductive and capacitive reactances, let's take a moment to consider impedance in a purely capacitive circuit, with just dc resistance and capacitive reactance, but no inductive reactance. The situation with a simple RC circuit is similar to that of a simple RL circuit. Capacitive reactance, like inductive reactance, is not a real resistance. It consumes no actual power. The ac signal flows in and out of a capacitor, but not through it.

In a capacitor, the current leads the voltage by 90 degrees (just the opposite situation of what happens in an inductor). This means the vector diagram of a dc resistance and pure capacitive reactance would be inverted from the resistance-inductive reactance diagram presented earlier. A typical vector diagram for a resistance-capacitive reactance combination is shown in Fig. 4-9. Notice that this is just a mirror image of the vector diagram we used in the purely inductive circuit.

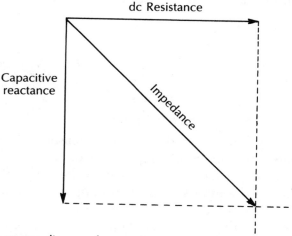

Fig. 4-9   *The vector diagram for a resistance-capacitive reactance combination is a mirror image of the vector diagram for a similar resistance-inductive reactance combination.*

Like an inductor, a practical capacitor has some degree of dc resistance. This internal dc resistance is effectively in parallel with the capacitive reactance. The internal dc resistance is normally a very high value—remember, a capacitor tends to oppose the flow of dc signals. Because the internal dc resistance is usually so much higher than the capacitive reactance, it doesn't have much effect on the parallel combination. Resistances in parallel always have a combined effective resistance lower than any of the individual component resistances. For this reason, the internal dc resistance of a capacitor can usually be ignored in most practical impedance equations, unless very high-precision applications are involved. Most real-world capacitors also exhibit a small (usually negligible) inductance, unless the signal frequency is very high.

Ignoring any inductive reactance component, combining a dc resistance with a capacitive reactance works in pretty much the same way as combining a dc resistance with an inductive reactance. For example, if we have a simple series RC circuit, like the one shown in Fig. 4-10, the impedance formula is

$$Z = \sqrt{R^2 + X_C^2}$$

Notice that this is the same formula used to solve for the impedance in RL series circuits. However, it is very important to remember that capacitive reactance is, by definition, always 180 degrees out of phase with inductive reactance.

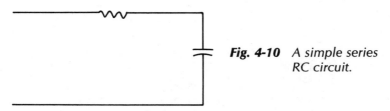

Fig. 4-10   *A simple series RC circuit.*

Also, since capacitive reactance decreases as the signal frequency increases, the circuit impedance of an RC series network must also decrease with an increase in the applied signal frequency. This is the opposite of what happens with an RL series circuit. Inductive reactance increases with an increase in the signal frequency, so the impedance of a series RL circuit must also increase with any increase in the applied signal frequency.

Impedance in inductive-capacitive circuits

Most practical electronic circuits include all of the basic passive electrical elements we have been discussing here—resistance, inductance, and capacitance. This doesn't necessarily mean that actual resistors, inductors, and capacitors will always be used. Other components also exhibit these various types of resistance.

Considering the crucial phase differences between inductive reactance and capacitive reactance, how do we go about finding the impedance of a circuit with a resistor, a capacitor, and an inductor in series, like the one shown in Fig. 4-11? Since the inductive reactance and the capacitive reactance are 180 degrees out of phase we obviously can't add them together.

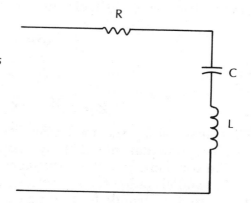

Fig. 4-11 *Practical electronic circuits almost always include dc resistances, capacitive reactances, and inductive reactances.*

Fortunately, this isn't as much of a problem as it might appear at first glance. When two similar signals (same waveshape) are 180 degrees out of phase with each other, they will directly oppose each other. Therefore, their combined value can be found by subtracting the smaller signal from the larger signal. This means we can combine the inductive reactance (X_L) with the capacitive reactance (X_C) to find the total effective reactance:

$$X_t = X_L - X_C$$

If the inductive reactance (X_L) is larger than the capacitive reactance, the total effective reactance will be positive. This means that the total effective reactance (X_t) will act like an inductive reactance. That is, the reactance will increase with an increase in the signal frequency.

On the other hand, if the capacitive reactance is larger than the inductive reactance, the total effective reactance will be negative, and act like a capacitive reactance. The reactance value will decrease as the signal frequency increases.

A special case occurs when the inductive reactance and the capacitive reactance are exactly equal. This condition is known as resonance, and it will be discussed in the next section of this chapter.

Since both inductive reactance and capacitive reactance change their values in opposite directions with changes in the signal frequency, it follows that the total effective reactance in a circuit with both a capacitance and an inductance will sometimes act inductively, and sometimes capacitively.

The total effective reactance (X_t) can then be combined with the dc resistance in the usual way to find the impedance of the circuit. For a series RCL circuit, the impedance is equal to

$$Z = \sqrt{R^2 + X_t^2}$$

or

$$Z = \sqrt{R^2 + (X_L - X_C)^2}$$

Of course, since the total effective reactance is squared, the resulting impedance will always have a positive value.

In some circuits, the inductance and the capacitance might be in parallel, rather than in series, as illustrated in Fig. 4-12. In this case, the formula for finding the total effective reactance is

$$X_t = \frac{(X_L)\,(X_C)}{(X_L - X_C)}$$

Again, if X_t turns out positive, the total effective reactance is inductive, but if X_t is negative, the reactance is effectively capacitive.

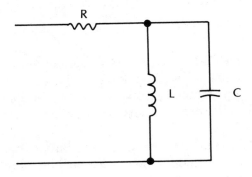

Fig. 4-12 *In some circuits, the inductance and capacitance might be in parallel rather than in series.*

Resonance

Capacitive reactance decreases as the applied frequency increases, while inductive reactance increases along with the applied frequency. Logically enough, at some specific frequency for every possible combination of a capacitor and an inductor, the capacitive reactance and the inductive reactance will be equal. This will only occur at one specific frequency for any particular combination of capacitor and inductor. This frequency is the resonant frequency, and it results in a noninductive circuit condition called resonance.

In a series LC circuit, the impedance equals

$$Z = \sqrt{R^2 + (X_L - X_C)^2}$$

If the inductive reactance and the capacitive reactance are exactly equal ($X_L = X_C$), these factors will cancel out, leaving

$$Z = \sqrt{R^2 + 0^2}$$

$$= \sqrt{R^2}$$

or simply

$$Z = R$$

At resonance, the impedance in a series LC circuit will be determined solely by the dc resistance in the circuit. Its impedance is at its minimum value at the resonant frequency.

Now, let's take a look at what happens at resonance in a parallel LC circuit. In this case, the total effective reactance equals

$$X_t = \frac{(X_L)\,(X_C)}{(X_L - X_C)}$$

When the inductive reactance and the capacitive reactance are equal ($X_L = X_C$), this becomes

$$X_t = \frac{(X_L)\,(X_C)}{0}$$

Any number divided by zero is always undefined and is usually considered nominally as infinity. Therefore, the nominal impedance in a parallel LC circuit at resonance is arbitrarily large. (In any practical electronic circuit, the actual impedance will be extremely high, but not truly infinite.) Clearly, a parallel LC circuit exhibits its maximum impedance value at the resonant frequency.

Any specific combination of an inductance and a capacitance will have one and only one resonant frequency. The formula for finding this single resonant frequency for a particular inductance-capacitance combination is:

$$F = \frac{1}{(2\pi\sqrt{LC})}$$

where

F = frequency, in hertz;
L = inductance, in henries; and
C = capacitance, in farads.

This same formula can be used for both series LC and parallel LC circuits.

As an example, let's suppose we have a 100-mH (0.1-H) coil and a 10-μF (0.00001-F) capacitor. The resonant frequency for this particular combination of components works out to approximately

$$F = \frac{1}{(6.28 \times \sqrt{0.1 \times 0.00001}\ )}$$

$$= \frac{1}{(6.28 \times \sqrt{0.000001})}$$

$$= \frac{1}{(6.28 \times 0.001)}$$

$$= \frac{1}{0.00628}$$

$$= 159.24 \text{ Hz}$$

This can be rounded off to 159 Hz.

In a practical electronic design situation, you'll usually need to approach this equation from a different angle. When designing a resonant circuit, you'll probably know the desired resonant frequency and will need to determine what component values to use to get the intended results. In this type of situation you can arbitrarily select a likely inductance value, and then rewrite the resonance equation to solve for the appropriate capacitance. The rearranged formula is

$$C = \frac{1}{(4\pi^2 F^2 L)}$$

The constant $4\pi^2$ has a value of approximately 39.5, so this equation can be rewritten as

$$C = \frac{1}{(39.5 F^2 L)}$$

As an example of how this formula works, let's say we need a circuit that is resonant at a frequency of 1000 Hz. We can arbitrarily select a 100-mH (0.1-H) coil to serve as the inductor. This means the necessary capacitance value is

$$C = \frac{1}{(39.5 \times 1000^2 \times 0.1)}$$

$$= \frac{1}{(39.5 \times 1,000,000 \times 0.1)}$$

$$= \frac{1}{3,950,000}$$

$$= 0.00000025 \text{ F}$$

$$= 0.25 \ \mu\text{F}$$

A 0.22-μF capacitor should do the job nicely.

Alternatively, we could start out with an arbitrarily selected capacitance and then solve for the necessary inductance. In this case, the formula is

$$L = \frac{1}{(39.5F^2C)}$$

Increasing either the inductance or the capacitance decreases the resonant frequency.

Coils

Now, let's look at inductors as physical electronic components. Inductors are generally called coils, and that is just what this type of component usually is—a coil of insulated wire wound around a core. The core can be made of powdered iron or some other magnetic material, or it may be nothing more than a small cardboard tube, or even just plain air. The construction of a typical coil is illustrated in Fig. 4-13.

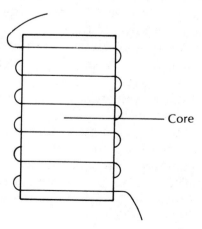

Fig. 4-13 *The construction of a typical coil.*

Core

The inductance of a coil is determined by a number of factors. The most important of these factors include the width of the core, the diameter of the wire, the number of turns of wire around the core, and the spacing between the individual turns of the coil. The material in the coil's core is also important. A core with low magnetic reluctance can increase the strength of the magnetic field generated around the coil, thereby increasing the voltage induced back into the turns of the coil.

Some coils are adjustable. That is, the inductance may be manually varied. Usually the core in a variable coil is constructed so that it can be moved slightly in and out of the center of the coil with a screw called a slug. This makes the core appear to be made partly of air and partly of (usually) powdered iron. Adjusting the proportion of air and powdered iron in the center of the coil varies the effective reluctance of the core and, thus, the inductance of the coil.

Many coils have the wire turns visibly exposed (but electrically insulated, of course), but some are sealed in metal cans to avoid undesired interaction with other components in the circuit. Without this shielding, the magnetic field could induce an unwanted current into nearby components and leads. Obviously, inducing a current where it's not intended can be quite detrimental to correct circuit operation.

The wire of a coil must normally be insulated, because the turns are generally wound quite closely together. If separate turns of uninsulated wire shift position and touch, allowing current to flow between them, a short circuit exists. This shorting of turns makes the coil appear to have fewer turns of wire as far as the current is concerned. This can change the coil's inductance, because one of the key factors in determining the inductance is the number of turns in the coil.

Figure 4-14 shows some of the most commonly used symbols for coils. A and B are fixed inductance coils, while the arrows through symbols C and D indicate that these coils are adjustable. Sometimes special symbols are used to indicate the core material of the coil. In this system, the six symbols in Fig. 4-14 are all air-core inductors. If two dotted lines are drawn beside the inductor symbol, as in Fig. 4-15, it means the coil's core is made of powdered iron. Two solid lines, as in Fig. 4-16, indicate that the coil's core is made of stacks of thin sheet iron.

However, most schematic diagrams simply use the standard symbols shown in Fig. 4-14, and the type of coil (and its core

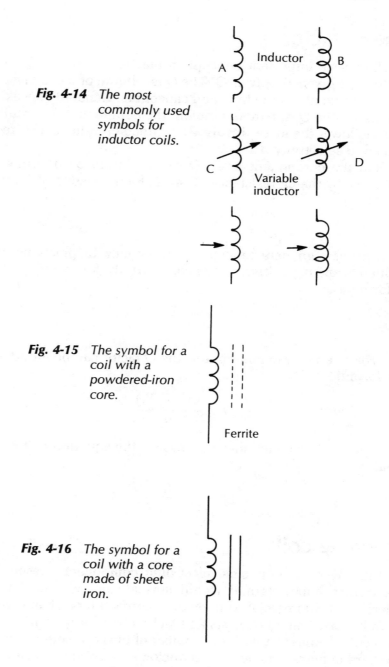

Fig. 4-14 *The most commonly used symbols for inductor coils.*

Inductor

A   B

C   D

Variable inductor

Fig. 4-15 *The symbol for a coil with a powdered-iron core.*

Ferrite

Fig. 4-16 *The symbol for a coil with a core made of sheet iron.*

material) is specified in the parts list. Coils and inductance values are usually represented by the letter L, since *I* is commonly used to represent current.

The Q factor

Another very important concept in electronics is the Q factor, sometimes given simply as Q. The Q is a figure of merit given for resonant circuits, inductors, and capacitors. Generally speaking, the higher the Q, the more efficient the component or circuit is, and the lower the losses involved. In tuned circuits, Q is directly related to selectivity.

On the component level, Q can be found by dividing the reactance by the dc resistance. That is, for a capacitor

$$Q = \frac{-X_C}{R}$$

The minus sign here reflects the difference in phase between inductive and capacitive reactance. Similarly, for an inductor the Q formula is

$$Q = \frac{X_L}{R}$$

For a series resonant circuit, the Q factor can be found with the formula:

$$Q = \frac{(X_C - X_L)}{R}$$

For a parallel resonant circuit, however, the equation is inverted. That is

$$Q = \frac{R}{(X_C - X_L)}$$

Winding coils

Unlike the case with most electronic components, resourceful electronics hobbyists and technicians are not restricted to the offerings of commercial component manufacturers when it comes to inductors. Many hobbyists and technicians learn to wind their own coils. Unless a very large number of turns is required, it isn't too hard to make a home-brew inductor, and if you work carefully, the result will work every bit as well as a commercially manufactured coil.

A number of factors determine the inductance of a coil. These include the core material and diameter, the number of turns in the coil, and how closely the turns are spaced. For a single-layer coil (one with no overlapping windings) wound on a nonmagnetic core, the formula for determining inductance is as follows:

$$L = \frac{(0.2d^2N^2)}{(3d + 9l)}$$

where

L = inductance of the finished coil, in millihenries;
d = diameter of the coil winding, in inches;
l = length of the coil winding, in inches; and
N = number of turns in the coil.

For example, let's say we are winding a coil on a 0.75-in. diameter nonmagnetic core. We will be using #32 enameled wire, and the coil will consist of 150 closely wound turns. The total length of the coil is 1.2 in. The inductance of this coil works out to

$$L = \frac{(0.2 \times 0.75^2 \times 150^2)}{(3 \times 0.75 + 9 \times 1.2)}$$

$$= \frac{(0.2 \times 0.5625 \times 22{,}500)}{(2.25 + 10.8)}$$

$$= \frac{2531.25}{13.05}$$

$$= 193.96552 \text{ mH}$$

If the number of turns (N) is increased, while the diameter (d) is held constant, the inductance (L) will be increased. The amount of this increase in inductance will depend on whether the coil's length (l) is increased by the added windings, or if this parameter is also held constant to the original value of l by squeezing the turns more tightly together.

Let's return to our earlier example and double the number of turns ($2N$). All of the other values, including length (l) will remain the same. That is, the relevant values this time are

- d = 0.75;
- l = 1.2; and
- N = 300 = (2 × 150).

How does this change affect the inductance of our coil? Let's use our formula again and find out.

$$L = \frac{(0.2 \times 0.75^2 \times 300^2)}{(3 \times 0.75 + 9 \times 1.2)}$$

$$= \frac{(0.2 \times 0.5625 \times 90{,}000)}{(2.25 + 10.8)}$$

$$= \frac{10{,}125}{13.05}$$

$$= 775.86207 \text{ mH}$$

Comparing this result to our earlier example, we find that the change in inductance from doubling the number of turns works out to

$$\frac{776}{194} = 4$$

Increasing the number of turns, while holding the diameter and coil length constant causes an increase in inductance that is equal to the square of the increase in the number of turns. In our example, we increased the number of turns by a factor of 2, so the inductance increased by a factor of

$$2^2 = 2 \times 2$$
$$= 4$$

Now, let's try the same problem, but this time we will assume that doubling the original number of windings also doubles the length of the coil. That is, the spacing between the windings is not changed this time. The relevant values for this example are

- $d = 0.75$;
- $l = 2.4 = (2 \times 1.2)$; and
- $N = 300 = (2 \times 150)$.

The inductance in this example works out to a value of

$$L = \frac{(0.2 \times 0.75^2 \times 300^2)}{(3 \times 0.75 + 9 \times 2.4)}$$

$$= \frac{(0.2 \times 0.5625 \times 90{,}000)}{(2.25 + 21.6)}$$

$$= \frac{10{,}125}{23.85}$$

$$= 424.5283 \text{ mH}$$

When increasing the number of turns (N) increases the coil length (*l*), while the diameter (*d*) remains constant, the original inductance value will be multiplied by a factor slightly greater than the multiple factor of *l* and *N*. In our example, *l* and *N* are increased by a factor of 2, and the inductance increases by a factor of

$$\frac{424}{194} = 2.185567$$

In practical applications, we will probably know the desired inductance value (*L*) and will need to determine how to wind a coil to achieve that specific value. The best way to do this is to select a likely core form of a given diameter (*d*) and arbitrarily pick out a reasonable length for the coil (*l*). Now the basic inductance formula can be algebraically rearranged to solve for the necessary number of turns.

$$N = \sqrt{\frac{(L(3d + 9l))}{0.2d^2}}$$

As an example, let's say we need a 100-mH coil and we happen to have a 1.2-in. diameter coil form. We'll set the coil length (*l*) at 1 in. This means the number of turns in our coil should be

$$N = \sqrt{\frac{(100\,(3 \times 1.2 + 9 \times 1)\,)}{(0.2 \times 1.2^2)}}$$

$$= \sqrt{\frac{(100\,(3.6 + 9)\,)}{(0.2 \times 1.44)}}$$

$$= \sqrt{\frac{(100 \times 12.6)}{0.288}}$$

$$= \sqrt{\frac{1260}{0.288}}$$

$$= \sqrt{4375}$$
$$= 66.14 \text{ turns}$$

We can round this off to 66 complete turns.

Several magnetic materials are often used for coil cores. These include such substances as iron, powdered iron, ferrite, and nickel alloy. When a magnetic core is used in an inductor, a crucial new factor enters into the equation when determining the inductance of the coil. This new factor is the permeability of the core material, and it is represented by the Greek letter μ (pronounced mu).

The equation for determining the inductance of a coil with a magnetic core is:

$$L = \frac{(4.06N^2\mu A)}{(0.27 \times 10^8 \times l)}$$

where

N = number of turns;
l = coil length, in inches;
A = cross-sectional area of the coil, in square inches; and
μ = permeability of the coil's core.

Let's try out a typical example, using the following values:

- N = 1000 turns;
- μ = 5000;
- A = 0.4 sq. in.; and
- l = 3 in.

The inductance of this particular coil works out to:

$$
\begin{aligned}
L &= \frac{(4.06 \times 1000^2 \times 5000 \times 0.4)}{(0.27 \times 10^8 \times 3)} \\
&= \frac{(4.06 \times 1{,}000{,}000 \times 5000 \times 0.4)}{810{,}000{,}000} \\
&= \frac{8{,}120{,}000{,}000}{810{,}000{,}000} \\
&= 10.024691 \text{ mH}
\end{aligned}
$$

To design a coil for a specific desired inductance value (L), you can solve for the number of turns (N) by rearranging the basic equation like this:

$$N = \sqrt{\frac{(0.27 \times 10^8 \times L \times l)}{4.06\mu A}}$$

or

$$N = \sqrt{\frac{(66,502,463.054 \times L \times l)}{\mu A}}$$

The standard coil core form is a straight cylinder. Another type of coil uses a toroidal core, as illustrated in Fig. 4-17. A toroid is a ring or doughnut-shaped object. This type of coil core offers a number of important advantages, including small size and compactness, high Q, and perhaps, most important of all, self-shielding. Toroidal coils can be operated at extremely high frequencies.

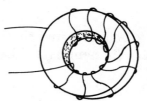

Fig. 4-17 *Some coils use a toroidal core.*

The formula for determining the inductance of a toroidal coil is as follows:

$$L = 0.011684N^2\mu h \times \log_{10}(OD/ID)$$

where

N = number of turns;
μ = permeability of the core material;
h = height of the core, in inches;
OD = outside diameter of the toroidal core, in inches; and
ID = inside diameter, in inches.

Manufacturers of toroidal cores often give specific inductances as a function of the number of turns. In such cases, there is no need to bother with the above calculations.

Magnetic cores for coils (both cylindrical and toroidal) are becoming increasingly difficult to find, especially in the electronics hobbyist market. They don't sell in large quantities these days, so most electronic parts suppliers don't bother to stock them. Fortunately, the electronics surplus houses are usually a good source for such items. Check out local surplus dealers in your area, or send for the catalogs from the mail-order dealers who advertise in the backs of popular magazines that cater to electronics hobbyists and technicians.

RL time constants

A circuit consisting of an inductor and a resistor in series, like
the one shown in Fig. 4-18, has a definite associated time constant, just as the resistance-capacitance (RC) circuits described in
chapter 3 do.

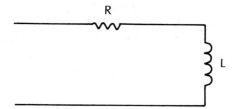

Fig. 4-18 *RL networks have
a time constant,
similar to RC
networks.*

For RL circuits, the time constant is found by dividing the
inductance (in henries) by the resistance (in ohms):

$$T = \frac{L}{R}$$

The time constant of an RL circuit is the time required for the
induced voltage to reach 63% of its full value.

For example, let's suppose we have a 100-mH (0.1-H) inductor in series with a 1000-Ω (1-kΩ) resistor. The time constant of
this combination is

$$T = \frac{0.1}{1000}$$
$$= 0.0001 \text{ s}$$
$$= 0.1 \text{ ms}$$

Increasing the inductance to 200 mH (0.2 H), while holding the
resistance constant changes the time constant to

$$T = \frac{0.2}{1000}$$
$$= 0.0002 \text{ s}$$
$$= 0.2 \text{ ms}$$

Increasing the inductance in an RL circuit increases the time
constant.

On the other hand, if we leave the inductance at 100 mH (0.1
H), but increase the resistance to 2000 Ω (2 kΩ), the time constant

is changed to

$$T = \frac{0.1}{2000}$$
$$= 0.00005 \text{ s}$$
$$= 0.05 \text{ ms}$$

Increasing the inductance in an RL circuit decreases the time constant.

Combining inductances

If a number of shielded coils are placed in series, as illustrated in Fig. 4-19, their inductance values will add, the same as with multiple resistances in series. That is,

$$L_t = L_1 + L_2 + L_3 + \ldots + L_n$$

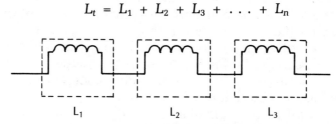

$$L_1 \qquad\qquad L_2 \qquad\qquad L_3$$

Fig. 4-19 *Shielded inductances in series add.*

If multiple shielded coils are placed in a parallel circuit, as shown in Fig. 4-20, the formula for finding the total effective inductance is also similar to the parallel resistance formula. The

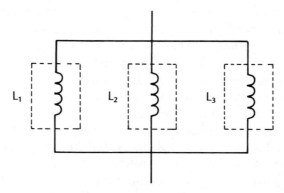

Fig. 4-20 *Shielded inductances in parallel use the same formula as dc resistances in parallel.*

reciprocal of the total effective inductance equals the sum of the reciprocals of each of the parallel inductances in the circuit. That is,

$$\frac{1}{L_t} = \frac{1}{L_1} + \frac{1}{L_2} + \frac{1}{L_3} + \ldots + \frac{1}{L_n}$$

You'll notice that I've specified shielded coils in these equations. That is because unshielded coils can interact. If multiple inductances are allowed to interact in the circuit, the equations are complicated by a factor called the coefficient of coupling.

Coefficient of coupling

If two unshielded coils are brought within fairly close proximity of each other, their magnetic fields will interact. The magnetic lines of force from one coil will cut across the turns of the other coil, causing the coils to each induce a voltage in the other, as well as the voltage they induce in themselves. This is called mutual inductance. So far, we have only considered self-inductance. The strength of the mutual inductance between coils is described as the coefficient of coupling.

If the coils are positioned so that only a few of their magnetic lines of force interact, they are said to be loosely coupled. If, on the other hand, the coils are placed very close to each other so that most of the magnetic lines of force from each coil cut across the other, the coils are closely coupled.

The mutual inductance can either aid or oppose the self-inductance of the coils, depending on the respective polarity of the magnetic fields. If each coil is wound in the opposite direction, both magnetic north poles or both magnetic south poles will face each other. Under these circumstances, the mutual inductance is in opposition to the self-inductance.

If, however, the coils are both wound in the same direction, we will have a magnetic north pole facing a magnetic south pole. The mutual inductance in this case will aid or add to the self-inductance. Mutual inductance is represented in electronics formulas by the letter M, and it is measured in henries (or millihenries), the same as ordinary self-inductance (L).

If two coils are in series and wound so that their mutual inductance aids their self-inductance, the total effective inductance of the pair can be found with the formula

$$L_t = L_1 + L_2 + 2M$$

Similarly, if the mutual inductance is such that the magnetic fields of the two coils oppose each other, the forumla is

$$L_t = L_1 + L_2 - 2M$$

The same basic principle works with two coils in parallel. If their magnetic fields aid each other, the formula is

$$L_t = \cfrac{1}{\cfrac{1}{(L_1 + M)} + \cfrac{1}{(L_2 + M)}}$$

If the magnetic fields of these two coils in parallel oppose each other, the formula becomes

$$L_t = \cfrac{1}{\cfrac{1}{(L_1 - M)} + \cfrac{1}{(L_2 - M)}}$$

The theoretical maximum amount of coupling between two coils is 100%. Obviously, this would occur when all of the magnetic lines of force of one coil cut across all of the turns of the other coil, and vice versa. In practice, 100% coupling can never be totally achieved. However, it is possible to come quite close to 100% coupling if both coils are wound on a single, shared core, as illustrated in Fig. 4-21. Toroidal coils are effectively self-shielding, because all of the magnetic lines of flux are contained in the core material.

Fig. 4-21 *Very tight coupling can be achieved by winding two coils on a common core.*

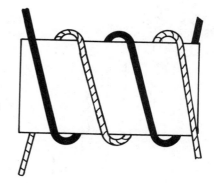

Transformer action

Interacting coils don't always need to be in the same electrical circuit. The magnetic field from the coil in one circuit can induce a voltage in the coil of a second, otherwise unconnected circuit. This process is called transformer action.

A transformer is a specialized type of inductor consisting of two (or sometimes more) coils wound on a single core and arranged so that their mutual inductance is at a maximum. The internal construction of a typical transformer is illustrated in Fig. 4-22.

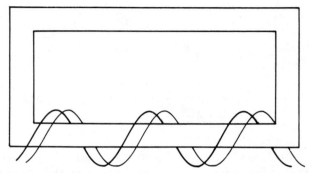

Fig. 4-22   *The internal construction of a typical transformer.*

Like coils, transformers are often classified according to their core material. Commonly used symbols for transformers are shown in Fig. 4-23. The symbol shown in Fig. 4-23A can be used

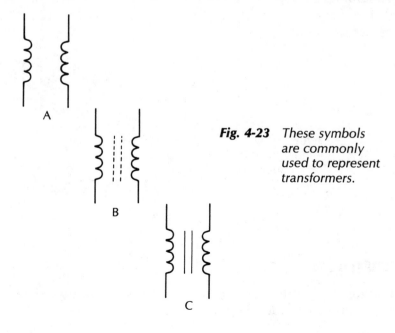

Fig. 4-23   *These symbols are commonly used to represent transformers.*

to represent any type of transformer or, specifically, an air-core transformer. Figure 4-23B represents a transformer with a powdered iron core. If the symbol shown in Fig. 4-23C is used, it indicates that the transformer in question has a core that consists of a stack of sheet iron wafers.

Let's examine what happens in an ideal, lossless transformer. (Of course this is a theoretical ideal, beyond the practical possibility of any real-world component.) In our discussion, we will be using the simple circuit shown in Fig. 4-24.

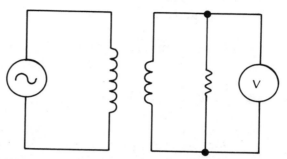

Fig. 4-24 *This simple circuit is used in the text to illustrate transformer action.*

Assuming that the ac voltage source puts out 100-V rms, the self-induced voltage in the first (or primary) coil will also be 100-V rms. Remember, we are assuming that there are no losses in this system. If the transformer's primary winding consists of 100 turns of wire, 1-V rms will be induced in each turn, so the total self-induced voltage of this coil is 100-V rms.

Assuming that the two coil windings in the transformer have 100% coupling, the same amount of magnetomotive force will cut across the turns of the other (or secondary) winding of the transformer. That is, 1 V will be induced in each turn of the secondary winding.

If the secondary winding also consists of 100 turns of wire, 100-V rms will be provided to the load circuit. This kind of transformer is called an isolation transformer, or a 1:1 transformer, because the turns ratio between the two windings is one to one. This type of device may seem rather useless, but it serves the important function of electrically isolating the load circuit from the power source. Isolation transformers are often used to reduce the risk of electrical shock or to prevent feedback.

However, in most transformers, the turns ratio between the windings is not one to one. What happens then?

For example, what if the primary winding still has 100 turns, but the secondary winding has only 25 turns? The magnetic field of the primary coil will still be the same, so 1-V rms will still be induced in each turn of the secondary coil. But, since there are only 25 turns of wire in this coil, the total voltage available to the load circuit is only 25-V rms. This kind of transformer is called a step-down transformer because the output voltage is stepped down or reduced from the input voltage. Incidentally, this explains the name "transformer." The voltage is transformed into a different value.

However, the same amount of current is induced in the secondary winding, regardless of the number of turns. Remember, current is always the same in all parts of a series circuit, so the current induced in one turn will flow through the entire coil. The power consumed by the secondary is the same as the power consumed by the primary. While the voltage is stepped down, the current is stepped up. The output (secondary) of the transformer puts out more current than its input (primary) takes in, but the voltage at the output is less than the voltage at the input.

Similarly, if the secondary winding of a transformer has more turns than the primary winding, the voltage will be stepped up (increased) and the current will be stepped down (decreased). This device, not surprisingly, is called a step-up transformer.

For example, if the secondary winding of our imaginary transformer had 250 turns to the primary winding's 100 turns, the output voltage across the secondary coil would be 250 V.

The 100-V rms voltage source and 100-turn primary coil were selected for these examples for simplicity. Not all practical transformers induce 1 V per turn. The exact voltage induced in each turn of the secondary winding depends on the source voltage and the number of turns in the primary winding. For instance, if 230 V were applied to the 100-turn primary winding of the transformer in our example, 2.3 V would be induced in each turn. If the primary coil consisted of 200 turns, the same 230-V input would induce 1.15 V per turn.

As you may suspect, changing the input voltage changes the output voltage too. Let's say we have a transformer that steps down a 100-V input to a 25-V output. The turns ratio is 4:1. If 230 V were applied to the input of this hypothetical transformer, the output voltage would be 57.5 V. (All of these voltages are rms.)

Practical transformers do not exhibit 100% coupling, of course, but they usually come close enough that we can round off a little and ignore the transformer losses, except in very critical applications. Transformer losses will be discussed in a later section of this chapter.

A step-down transformer sometimes can be used as a step-up transformer simply by reversing the connections to the windings. That is, use the original primary as the secondary and the old secondary as the primary. The transformer itself doesn't care which is which. If a voltage is applied to one winding, that winding will serve as the primary and induce a voltage in the other winding. However, depending on the internal design of the transformer, this might not always work. It could result in blown fuses or other risks. Be careful.

Center taps

Many transformers have an extra connection point, or tap, at the midpoint of the secondary winding. The symbol for such a center-tapped transformer is shown in Fig. 4-25. The secondary winding of a center-tapped transformer can be considered as two coils in series connected at the center tap. The center tap on an autotransformer is often manually moveable, making it an adjustable transformer.

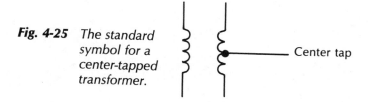

Fig. 4-25   *The standard symbol for a center-tapped transformer.*   Center tap

As an example of how a center-tapped transformer works, let's assume that the secondary winding consists of 100 turns of wire. If the center tap (point B) is at the exact midpoint of this coil, there will be 50 turns of wire on either side of the tap. If the primary induces 1 V per turn in the secondary, the full output between points A and C is 100-V rms (100 turns × 1 V per turn).

From point A to point B there are only 50 turns. With 50 turns at 1 V per turn, this means there are 50 V between A and B. Similarly, there will be 50 V across the 50-turn half-coil between points B and C. Notice that the two half-winding voltages equal the full secondary winding voltage. That is,

$$AB + BC = AC$$

If the center tap is grounded, as in Fig. 4-26, the voltage across AB and AC will look like the graphs shown in Fig. 4-27. Notice that these two voltages are equal (50-V rms with respect to ground—the center tap), but they are 180 degrees out of phase. When voltage AB goes up, voltage BC goes down, and vice versa. As you can see, a center-tapped transformer can be very helpful in the design of a dual-polarity power supply circuit.

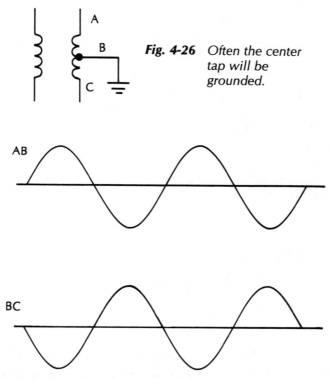

Fig. 4-26 Often the center tap will be grounded.

Fig. 4-27 Grounding the center tap produces two voltages that are 180 degrees out of phase with each other.

Autotransformers

An autotransformer is a unique form of transformer that uses just a single coil winding as both the primary and the secondary.

This is accomplished by using the center-tap principle. The basic construction of an autotransformer is illustrated in Fig. 4-28. The symbol used to represent an autotransformer is shown in Fig. 4-29.

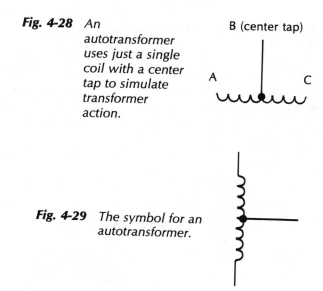

Fig. 4-28 *An autotransformer uses just a single coil with a center tap to simulate transformer action.*

Fig. 4-29 *The symbol for an autotransformer.*

Let's say there are 100 turns in the entire coil, and 100-V rms is applied between the extreme ends of the winding—points A and C. Because the entire voltage is distributed equally across the entire coil, the self-induction will equal 1-V rms per turn.

Now, let's suppose the center tap (B) is placed so that there are 70 turns of wire between A and B and only 30 turns from B to C. If we took the voltage across BC, we'd have a 30-turn coil with 1-V rms induced across each turn. The output voltage would therefore be 30-V rms. The single coil acts like two very closely placed but separate coils. In other words, we get transformer action from a single coil.

An autotransformer cannot be used for electrical isolation applications. Because the primary and secondary windings are both the same coil, there can be no electrical isolation between them.

We have already seen how an autotransformer can be used as a step-down transformer. To use an autotransformer as a step-up transformer, apply the input voltage between points B and C and tap off the output voltage across points A and C.

However, an autotransformer cannot always be used in place of a regular step-down or step-up transformer. For most practical transformer applications, some degree of electrical isolation is required, so the standard two-coil transformer must be used.

Other types of transformers

Some transformers have more than one secondary winding. An example of this is shown in Fig. 4-30. The primary coil in this transformer is AB. Coil AB induces a voltage into both coil CD and coil EF. These two coils are both secondaries. The circuits connected to these two secondaries can be completely isolated from each other electrically.

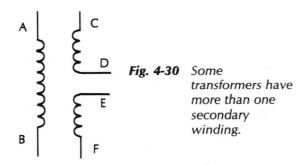

Fig. 4-30 *Some transformers have more than one secondary winding.*

So far, we have been concentrating solely on power transformers, which usually (except for the isolation transformer) change the voltage (either up or down). Transformers are also widely used for impedance matching. That is, the reactance of the primary winding is different from the reactance of the secondary winding.

In an ac circuit, the best and most efficient power transfer occurs when the input and output impedances match. A transformer can be used to create such an impedance match where it wouldn't otherwise exist.

There are usually some design differences between power transformers and impedance-matching transformers. If a transformer is rated in ohms, it is an impedance-matching transformer. If it is rated in terms of its voltage, it is a power transformer. Most power transformers in the United States are rated assuming an input voltage of 120-V rms. This is the voltage available from normal ac power lines in the home, so it is a very reasonable and practical standard.

Losses in a transformer

So far we have been dealing with an ideal, lossless transformer that wastes no energy. All of the energy put into the primary coil is induced into the secondary winding. In actual practice, of course, such perfection is not possible. You'll always have to put more power into a transformer (or any other electronic component, for that matter) than you can get out of it. This is because the transformer (or other resistor component) will use up or "lose" some of the electrical energy. In any real-world component, there will be some dc resistance. That is, some of the electrical energy will be converted into heat and dissipated. Also, the two coils in a transformer can never achieve 100% mutual inductance. Some of the magnetic energy will not cut across any of the coils of the secondary winding, and that energy is lost as far as the transformer is concerned.

Most of the losses in a practical transformer can be made quite small with good design. Generally, transformer losses can be ignored, especially if you are only interested in rough, ballpark figures. However, even these minor losses can be significant in some cases.

For instance, let's assume we have a transformer with 100 turns in each winding (a 1:1 transformer). If we apply 100-V rms across the primary winding of this isolation transformer, we should get a 100-V output across the secondary winding. In an actual circuit with a practical transformer, the output might only be about 98.5-V rms. Another transformer with the same basic specifications, but greater internal losses, might only put out 79-V rms.

In most cases, commercially available power transformers are marked with their intended input and output voltages. These figures usually take most internal losses into account. You will rarely have to be concerned with the actual turns ratio of a transformer (unless, of course, you are making your own), just the input:output voltage ratio. You should, however, have some understanding of the major causes of energy losses in transformers.

One important limitation has already been mentioned. True 100% mutual inductance can never be achieved between any two practical coils, no matter how closely they are placed. Some of the magnetic lines of force will inevitably be lost into the surrounding air. However, it is quite possible to achieve a very high coefficient of coupling in a transformer. Manufacturers of transformers can also compensate for the difference due to lost magnetic energy by adding a few extra turns to the secondary winding.

Since 100% coupling cannot be achieved, some of the magnetic field is effectively leaked out of the circuit. These lost magnetic lines of force are called leakage flux, and the effect is called leakage reactance. Leakage flux in a transformer can be greatly reduced by winding the coils on an iron core or a core made of some other low-reluctance substance.

The shape of the transformer's core can also have a significant effect on the leakage reactance. Figure 4-31 shows some of the most commonly used core shapes in modern transformers. The shell core (Fig. 4-31C) typically has the lowest leakage reactance, but it is also the most expensive type of core to manufacture.

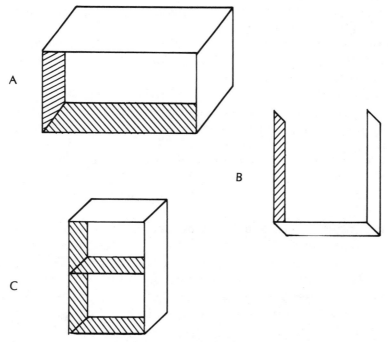

Fig. 4-31 *Various core shapes are used in transformers.*

Another important source of energy loss in a transformer is the dc resistance. Any practical conductor always has some degree of resistance, which inevitably increases as the length of the conductor increases. As far as a dc voltage is concerned, a coil is just a very long wire, wound into a small space. It acts like a very low-value resistor. This dc resistance is usually very small— almost never as much as 100 Ω. But even though this dc resistance is small, it does result in a small voltage loss. In a transformer, the dc resistance is sometimes referred to as copper loss, because the coils themselves are usually made of copper wire.

The dc resistance between the separate windings of a transformer (from the primary to the secondary) should be virtually infinite because there is no direct electrical connection between them. They are only connected magnetically. (Of course, the autotransformer is an exception to this rule.) If you find a measurable resistance between the primary and secondary windings, the transformer is probably shorted and should be discarded. Not only will it not work properly, but it could be a fire hazard.

We must also consider the ac inductive reactance of the transformer's coils. Resistances and reactances, as you should recall, cannot simply be added together. Impedances and reactances were discussed at length earlier in this chapter. Remember that inductive reactance increases as the applied signal frequency increases. Since the coils are conductors and they are separated by insulation (the insulation on the wire, the surrounding air, and the core), unwanted capacitances can appear within the transformer, contributing to the power loss.

Another important type of energy loss in a transformer is iron loss. There are actually two forms of iron loss. The first is known as hysteresis loss. Since the iron core is within a strong magnetic field (generated by the current flowing through the surrounding coil), the core becomes magnetized. As the alternating current flows through the coil, the magnetic charge of the core is forced to continually alternate its polarity. This process uses up a certain amount of energy because the iron core will oppose any change in its condition. This energy must be "stolen" from the electrical current because that is the only energy source within the system. In other words, some power is lost due to hysteresis loss. Hysteresis loss can be reduced by constructing the transformer's core of some material that is very easy to magnetize and demagnetize. Some materials of this type are silicon steel and certain other alloys.

The other form of iron loss in a transformer is due to the electrical current that is induced in the conductive core by the fluctuating magnetic fields of the coils. This current is called an eddy current, and it represents wasted power that is not allowed to reach the secondary winding and the desired load circuit.

To reduce eddy currents as much as possible, transformer cores are often made up of a pile of very thin metallic sheets (called laminations) rather than a solid mass of iron or whatever alloy is being used. These laminations are individually insulated from each other by a layer of varnish or oxide, so eddy currents cannot flow through the entire core and increase in strength.

Another inevitable source of power loss in transformers is called reflected impedance. As current flows through the primary winding, a magnetic field is created. This magnetic field induces a voltage in the secondary winding. So far, this is simply the desired transformer action. But because there is now a current flowing through the secondary coil, it also generates a magnetic field of its own, that induces an interfering current back into the primary winding. The energy for this back-induced current is taken from the secondary winding's circuit (the load).

Despite all these various losses, commercially available transformers are generally surprisingly efficient. Tolerance figures usually aren't given for transformer ratings. When in doubt, it is a simple matter to measure the actual output voltage across the secondary winding while feeding in a known source voltage across the primary winding. Remember, this will always be an ac voltage. Use the appropriate setting on your multimeter.

Motors

Closely related to the basic inductor and transformer is the motor. I decided to include this type of device in this chapter, but it could have fit into chapter 9 of this book, because a motor really is a type of transducer. A transducer converts one type of energy into another. In electronics work, one of these types of energy is always electrical energy. A motor converts electrical energy into mechanical energy. That is, an electrical signal through a motor can cause something to physically move. A motor is a practical application of the electromagnetic field surrounding a coil when current passes through it.

There are many different types of motors. Some are extremely tiny, while others are quite huge. Some small motors can only move very small loads, while some large motors can move tons. Some motors are designed to run on a dc voltage, and others use an ac voltage as the power source. Regardless of these differences, all motors are basically the same, at least in their fundamental operating principles.

A motor has two main parts—a movable coil and a fixed-position permanent magnet or a second, fixed-position coil. An electrical current is fed through a set of coils, setting up a strong magnetic field. The attraction of opposite magnetic poles and the repulsion of like magnetic poles results in the mechanical motion of the motor.

Figure 4-32 shows a simplified cut-away diagram of a typical motor. Notice that there are two sets of coils in this motor. One is stationary and is known as the field coil. The other coil, which is called the armature coil, or simply the armature, can freely rotate within the magnetic field of the field coil. The motor's shaft is connected directly to the movable armature coil. As the armature coil moves, the motor shaft rotates.

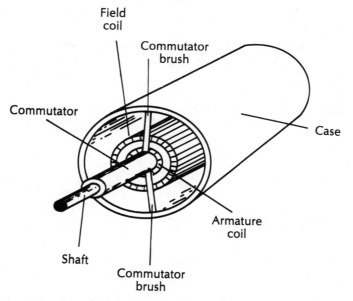

Fig. 4-32 *A simplified cut-away diagram of a typical motor.*

The commutator reverses the polarity of the current with each half-rotation of the armature and the shaft. This keeps the armature coil in constant motion, moving its magnetic poles from the like magnetic poles of the field coil.

The operation of a motor is illustrated in Fig. 4-33. In Fig. 4-33A, the armature coil is positioned so that its motor poles are

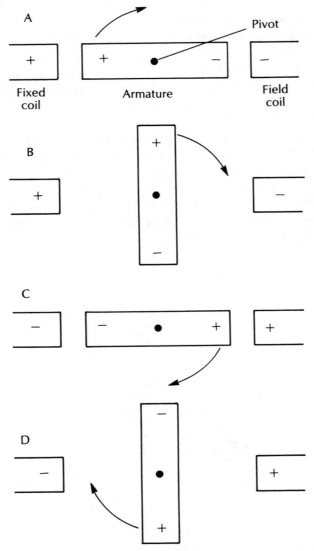

Fig. 4-33 *The changing magnetic poles force the armature coil in a motor to rotate.*

lined up with the like poles of the field coil. These like magnetic poles repel each other, forcing the armature coil to rotate, as shown in Fig. 4-33B. At some point, the attraction of the unlike magnetic poles will take over, pulling the armature into the position shown in Fig. 4 -33C.

The commutator now reverses the polarity of the current, so once again the like magnetic poles of the armature coil and the field coil are lined up, repelling each other. The whole process repeats for the second half-rotation, as illustrated in Fig. 4-33D, bringing us back to the original position of Fig. 4-33A. The commutator reverses the current polarity again and a new cycle begins.

Assuming that all other factors remain equal, increasing the current through the coils will increase the torque of the motor. That is, a motor can turn a larger (heavier) load if a larger current is supplied to the motor's coils. Another way of looking at this is to say that using a given motor to move a load, the heavier the load is, the more current the motor will be forced to draw from the power supply.

Some motors are designed to operate at a constant speed. Other motors are designed to change their rotation speed with changes in the applied current or voltage. The size of the load can also affect a motor's rotation speed. Obviously, heavier load weights will tend to slow the motor down because it has to work harder to turn the load. This is especially true if a constant current source is driving the motor. Some motors are intentionally designed so the load mass controls the actual rotation speed, within specific limits.

Generally, dc motors and ac motors are not interchangeable. Using the wrong type of power source can damage or destroy the motor. Excessive loads can also damage small motors under some operating conditions.

Like coils and transformers, motors can become defective if there is a short in one of the coil windings. Occasionally, a broken wire in one of the motor's coils could result in an open circuit condition. Of course, either an open motor or a shorted motor will not operate. When using motors in electronic circuits, remember they are inductive devices, and as such they affect the impedance of the circuit.

A special type of motor is the stepper motor. This type of motor consists of two (or more) fixed-position coils and a pivoted permanent magnet, that is free to rotate to align its magnetic poles with the unlike magnetic poles of the two coils. This rotating permanent magnet is called the rotor. A simplified diagram of a stepper motor is shown in Fig. 4-34.

In use, one coil of the stepper motor is energized at a time. This causes the rotor to move into any of four discrete positions, spaced 90 degrees apart. If both coils are simultaneously energized, we get four more discrete positions for the rotor, each 45

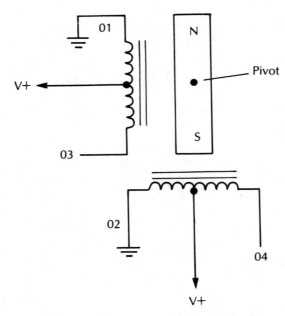

Fig. 4-34 *Stepper motors are used to position the load at a specific angle.*

degrees away from the first set. In other words, the rotor has eight discrete positions, or steps, per rotation cycle, as illustrated in Fig. 4-35.

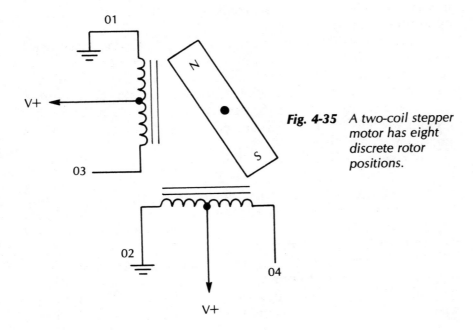

Fig. 4-35 *A two-coil stepper motor has eight discrete rotor positions.*

Most practical stepper motors have more than two sets of fixed coils, offering a wider variety of possible rotor positions, or steps. A typical practical stepper motor has a step angle of 1.8 degrees. This means that there are 200 possible step positions for the rotor.

Stepper motors are used for precise positioning of a load. A typical application is to adjust the directional position of an outdoor television or radio antenna from inside the house.

❖ Part 2
Active components

❖ 5
Diodes

SO FAR WE HAVE ONLY CONSIDERED PASSIVE COMPONENTS. FEW practical electronic circuits consist of passive components only. Most also include one or more active components. An active component, unlike a passive component, is capable of amplification. That is, a passive component can only reduce the amplitude of the signal, but an active component has gain and can therefore boost the signal's amplitude.

In the vast majority of modern electronic circuits, active components are always solid-state devices. They are made of a class of materials known as semiconductors. Vacuum tubes were common in older electronic equipment, but today they are virtually obsolete except in a few odd applications. One exception is the CRT (cathode-ray tube), which is still widely used in television sets, computer monitors, and oscilloscopes. Even in these specialized applications, it is highly probable that in the next decade the LCD (liquid-crystal display) will become the norm, and the tube will become truly obsolete. (LCDs will be covered in chapter 9.)

Semiconductor devices can be divided into three broad classes: diodes, transistors, and integrated circuits (ICs). Integrated circuits, in turn, can be classified as either linear or digital. Each of these semiconductor classifications will be covered in the next four chapters.

In this chapter we will be considering the diode. The name comes from *di*, meaning "two," and *ode* which is a shortened form of "electrode." In other words, a diode is a component with two electrodes. This terminology is really a holdover from the days of vacuum tubes, but it is still applied to the solid-state equivalent of the old diode tube.

Actually, a diode is not a truly active component because it is not capable of gain. However, it is much closer to the active transistor than to the passive resistor, capacitor, or inductor. Before discussing the diode, we should pause for a moment to consider semiconductors in general.

Semiconductors

Most materials can be considered as either conductors or insulators. A conductor permits electrical current to freely pass through it, while an insulator blocks the flow of electrical current.

No practical conductor is perfect. Not all of the current will get through it. Similarly, all practical insulators are less than perfect. Some electrical current can pass through even the best insulator. This is why it is dangerous to touch power lines. The cables are insulated, but the current flowing through the power lines is so high, a dangerous level of current can still get through the layers of insulation.

A conductor is a substance with a very low electrical resistance, and an insulator is a material with a very high electrical resistance. Some conductors are better (have a lower resistance) than others, and some insulators do a better job (have a higher resistance) than others. Not surprisingly, not all known substances fall neatly into the conductor or insulator categories. Some materials have a moderate resistance—too high to be a good conductor, but too low to serve as an insulator. Such substances are called semiconductors.

Two of the most commonly used semiconductors are germanium and silicon. For the moment, we'll confine our discussion to germanium, but remember that the same basic principles also apply to silicon (and other semiconductor materials).

If you are familiar with basic atomic physics, you probably know that electrons orbit around the nucleus of an atom, as illustrated in Fig. 5-1. The electrons are arranged in layers, or rings (sometimes called shells). The electrons in the outermost ring of any atom are called valence electrons. A germanium atom has four valence electrons. This is illustrated in Fig. 5-2.

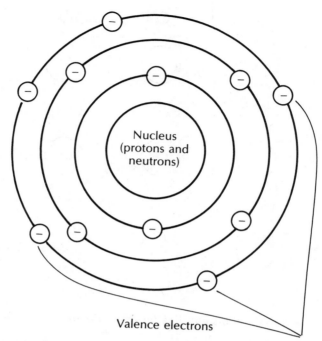

Fig 5-1   *The electrons in the outermost layer around the nucleus of an atom are called valence electrons.*

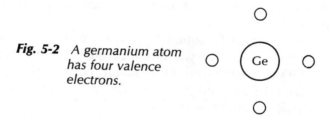

Fig. 5-2   *A germanium atom has four valence electrons.*

Ordinarily, the valence electrons in germanium pair up with the valence electrons of other, adjacent germanium atoms in a crystalline structure. The atoms are linked by sharing pairs of electrons, as shown in Fig. 5-3. The atoms within the crystal are held together by a force called the covalent bond. This name comes from the fact that the atoms are sharing their valence electrons.

Beyond being a fair resistor, pure germanium has no particularly unique or interesting electrical properties. By itself, it

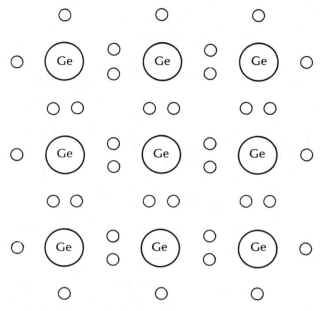

Fig. 5-3 *In a crystal, atoms interlink by sharing pairs of valence electrons.*

wouldn't be of much use in electronics. But, if selected impurities are added to a germanium crystal, a number of interesting and useful effects can be achieved. The process of adding specific impurities to a piece of pure semiconductor material is called doping.

One type of impurity sometimes used to dope germanium crystals is arsenic. Only a very, very small amount of the impurity is added to the pure semiconductor crystal. For simplicity in our discussion, we will assume just a single arsenic atom has been added to a small germanium crystal.

The impurity (the arsenic atom) will attempt to mimic a germanium atom in the crystalline standard. But arsenic has five valence electrons, while germanium has just four. When the arsenic atom is placed in the germanium crystal, there is an extra, unpaired electron, as shown in Fig. 5-4. This spare electron can drift freely throughout the crystal from atom to atom. In effect, we could say this electron is "lonely" and is "looking for a partner." But it can't find one.

This gives a small negative electrical charge (the extra electron) drifting aimlessly about within the crystal. The crystal as a whole is electrically neutral because the total number of protons equals the total number of electrons. If a germanium crystal is

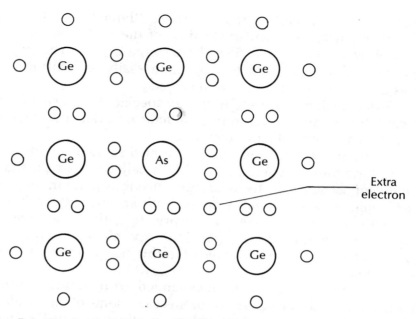

Fig. 5-4 *Doping a germanium crystal with arsenic leaves extra valence electrons.*

doped with a number of arsenic atoms, there will be a number of surplus electrons drifting through the crystal. The crystal itself will still be electrically neutral, of course. If a voltage source is connected to the doped crystal, as shown in Fig. 5-5, the excess electrons will be drawn to the positive terminal of the voltage supply and removed from the crystal.

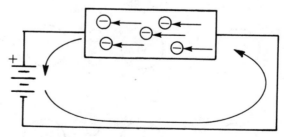

Fig. 5-5 *Current can flow through a doped semiconductor crystal.*

Because the crystal now has fewer electrons than it has protons, it possesses a positive electrical charge and draws more electrons out of the negative terminal of the voltage source. These electrons will move through the crystal without finding a place to latch into the crystalline matrix. These "homeless" electrons will flow out to the positive terminal of the voltage source. In other words, current will flow through the crystal. The doped crystal will allow current to flow more easily (exhibiting less resistance) than a pure germanium crystal.

So far we don't have anything very special. All we've done is lower the resistance of the semiconductor material. But there is more to the semiconductor than this.

There are two different types of doped semiconductors. So far we have only discussed one. This is called an N-type semiconductor because negatively charged electrons move through it. The semiconductor crystal is doped with an impurity with one valence electron too many. Not surprisingly, the other type of doped semiconductor uses an impurity element with one valence electron too few. The result in this case is a P-type semiconductor.

Germanium crystals are often doped with indium, which has just three valence electrons. In this case, some of the covalent bonds in the crystal are incomplete, as illustrated in Fig. 5-6.

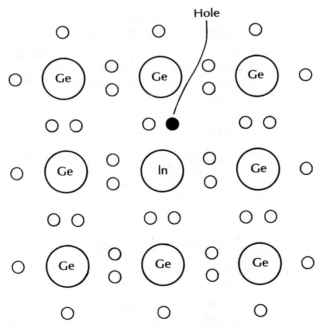

Fig. 5-6 *Doping a germanium crystal with indium leaves "holes" for missing valence electrons.*

There are empty holes where electrons belong, but there aren't enough electrons to fill all the holes.

The various covalent bonds in the crystal will "steal" electrons from one another to fill their holes. In effect, the positions of the holes appear to drift through the crystal. We can say we have a flow of holes. Actually, electrons are still being moved about, as in any electrical current, but in this situation, it is easier and more convenient to think of the holes as moving. Remember, a hole is simply the absence of an electron. We can think of the holes as positively charged pseudoparticles, since subtracting an electron will leave a positive charge. This allows us to greatly simplify our discussion of semiconductor action.

If the impurity used to dope the semiconductor crystal adds extra electrons (like arsenic), it is called a donor impurity. Other commonly used donor impurities are antimony and phosphorus. On the other hand, a doping impurity (such as indium) that adds extra holes (fewer electrons) is known as an acceptor impurity. Aluminum and gallium are also frequently employed as acceptor impurities.

An electric current will flow through either an N-type or a P-type semiconductor. With an N-type semiconductor we usually speak of a flow of electrons (as with ordinary electrical current), but with a P-type semiconductor we more frequently consider the flow of holes. Remember that the flow of holes moves in the opposite direction as the flow of electrons. Really, both flows take place simultaneously. The flow of electrons moves from the voltage source's negative terminal to the positive terminal. The flow of holes, on the other hand, starts at the voltage source's positive terminal and moves to the negative terminal.

Electrons and holes in a semiconductor crystal are referred to as current carriers, or simply carriers. Both types of doped semiconductors contain both types of carriers, but one kind of carrier will be more plentiful than the other. In an N-type semiconductor, for example, electrons are the major carriers and holes are the minor carriers. That is, there are more spare electrons than spare holes. The situation is exactly reversed with a P-type semiconductor, of course. Here, the holes are the major carriers and the electrons are the minor carriers.

Neither type of doped semiconductor exhibits any special electrical properties on its own. But when an N-type semiconductor is combined with a P-type semiconductor, some very special effects come into play.

The *PN* junction

In practical electronic components, a slab of N-type semiconductor is combined with a slab of P-type semiconductor. The point at which the different types of semiconductors are joined is called a junction, or, more precisely, a PN junction. When no external voltage is applied across a PN junction, the carriers (electrons and holes) are randomly placed throughout the two-part crystal, as illustrated in Fig. 5-7. Of course, the electrons will tend to be concentrated in the N-type section, but there will be some loose electrons (minor carriers) in the P-type section. Similarly, most of the holes will be in the P-type section, but the N-type section will also contain a few holes.

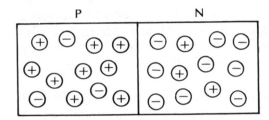

Fig. 5-7 *In electronic components, N-type and P-type semiconductors are combined into a PN junction.*

Remember that these extra electrons and holes represent irregularities in the crystalline structure of the doped semiconductors. In terms of the atoms in the semiconductor, the number of electrons is equal to the number of protons, so each section of the semiconductor is electrically neutral.

Now, suppose we hook up a dc voltage source so that its positive terminal is connected to the N-type semiconductor and the negative terminal is attached to the P-type semiconductor. This situation is illustrated in Fig. 5-8. As a result of the applied voltage, the holes in the P-type semiconductor will be drawn towards the edge of the crystal with the most negative charge. That is, the holes in the P-type section will move away from the junction. At the same time, the free electrons in the N-type semiconductor will be drawn away from the junction toward the positive terminal of the voltage source.

As you can see, the major carriers in each half of the semiconductor will be drawn to the outside edges of the crystal. There will be virtually no major carriers near the PN junction. This means that almost no electrons can cross the junction from one type of semiconductor to the other. Almost no current will flow

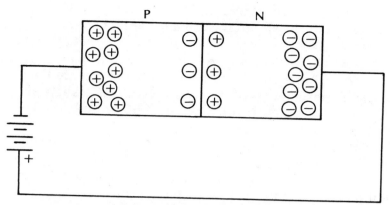

Fig. 5-8 *A reverse-biased* PN *junction.*

through the crystal. (There will be some slight current flow due to the minor carriers, but this current will be very small.) Under these conditions, the PN junction is said to be reverse biased. The semiconductor as a whole will exhibit a very high resistance.

If we reverse the polarity of the voltage source, as shown in Fig. 5-9, we have a very different situation. This time, the voltage source's negative terminal is connected to the N-type semiconductor and the positive terminal is connected to the P-type semiconductor. The positive charge on the P-type section will attract its minor carriers (electrons). Some of these electrons will leave the semiconductor crystal and flow into the voltage source's positive terminal. Since some electrons have been removed from the P-type section, but the number of protons remains unchanged, this half of the semiconductor crystal now has an overall positive electrical charge. In addition to drawing out the minor carriers

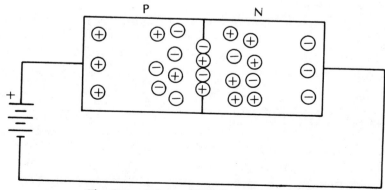

Fig. 5-9 *A forward-biased* PN *junction.*

(electrons), the positive terminal of the voltage source repels the major carriers (holes), forcing them towards the junction.

Meanwhile, the negative terminal of the voltage source is connected to the slab of N-type material, repelling its major carriers (electrons) towards the junction. Since there are many electrons being pushed towards the junction, and a positive electrical charge is pulling them from the other side, a number of electrons are forced through the narrow junction area to attempt to neutralize the positively charged P-type section. The negative terminal of the voltage source replaces the electrons that jump the junction with new electrons, which are also repelled towards the junction and pulled across. At the same time, the voltage source is continually drawing away more electrons from the P-type side, so it retains its positive electrical charge.

Clearly, current flows through the semiconductor crystal under these conditions. When a voltage is applied to a *PN* junction with this polarity, permitting current to flow, the junction is said to be forward biased. A forward-biased *PN* junction exhibits a very low resistance.

You can also look at the whole procedure from the point of view of a flow of holes, if you prefer. The voltage supply's negative terminal connected to the N-type material adds electrons to this section of the semiconductor crystal. These added electrons fill the holes (minor carriers) in the N-type section. Since there are now more electrons than in the normal, neutral state, this section acquires a negative electrical charge. This pulls fresh holes from the P-type side (where they are the major carriers). These P-type side holes are also pushed towards the junction by the positive terminal of the voltage source. In both cases we are describing the exact same phenomenon. All that has changed is our point of reference.

In summary, a PN junction permits one-way current flow. It exhibits a very low resistance and a high current flow when forward biased, but a very high resistance and little or no current flow when reverse biased.

The junction diode

A semiconductor diode is nothing more than a *PN* junction enclosed in a protective housing, usually plastic or glass, although metal cases are occasionally used. Connecting leads are attached to either end of the component. The technical name for this type of diode is a junction diode. As we shall discover later in this chapter, there are other types of diodes too. Usually, when technical literature refers to a diode without specifying the type, a junction diode can be assumed. However, there are some exceptions, so be careful.

The standard symbol for a junction diode is shown in Fig. 5-10. Some technicians add a ring around the diode symbol, as illustrated in Fig. 5-11, but this is usually considered redundant and unnecessary. These two symbols mean exactly the same thing and can be considered interchangeable. Don't be thrown if you run across a schematic that uses the alternate form. Also, this diode symbol is sometimes used to represent a tube-type diode, although this is extremely uncommon in modern electronic equipment.

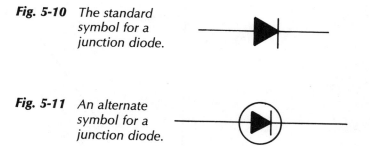

Fig. 5-10 *The standard symbol for a junction diode.*

Fig. 5-11 *An alternate symbol for a junction diode.*

The two leads of a diode are called the anode (positive lead) and the cathode (negative lead). The arrow in the diode symbol points to the cathode. This arrow indicates the flow of holes through the PN junction (when it is forward biased). Current flow (the flow of electrons) runs in the opposite direction from this arrow. That is, when the diode is forward biased, current flows from the cathode to the anode.

The case of a semiconductor diode is usually marked to indicate which end is the cathode. Some diodes have a band painted around their housing close to the end with the cathode lead, as shown in Fig. 5-12. Other diode casings are tapered at the cathode end, as illustrated in Fig. 5-13.

The operation of a typical junction diode is shown in Fig. 5-14. To the left (negative side) of the 0-V line, the diode is reverse biased. To the right (positive side) of this line, the diode

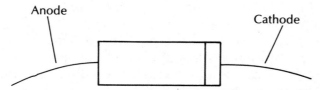

Fig. 5-12   *Some diodes have a band painted on the end close to the cathode lead.*

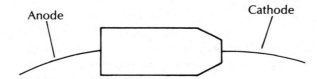

Fig. 5-13   *The body of some diodes is tapered at the cathode end.*

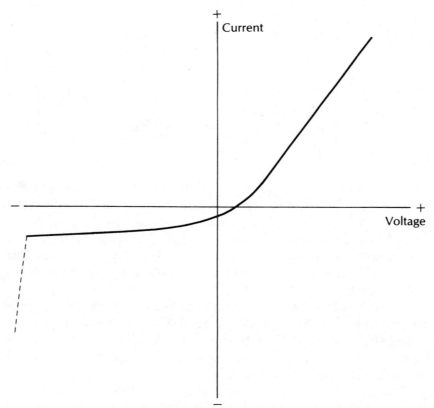

Fig. 5-14   *The operation of a typical junction diode.*

is forward biased. Notice that when it is reverse biased, the current flow through the diode remains fairly stable until the point where the reverse breakdown voltage is exceeded. At that point the negative current flow begins to increase very, very rapidly and the diode can quickly destroy itself with the extreme heat dissipation caused by the excessive current flow. When only slightly reverse biased, the diode has a narrow region where the current flow is building up to its stable, nominal reverse-biased value.

When the diode is forward biased (the applied voltage is positive), it has a small region where the current builds up fairly slowly until the applied voltage exceeds the normal forward voltage drop of the diode. For a silicon diode, this voltage drop is usually between 0.6 V and 0.8 V. Germanium diodes typically have lower forward voltage drops—usually 0.1 V to 0.3 V.

Once this forward voltage drop point has been exceeded, the current through a diode increases extremely rapidly. Once it is truly forward biased, the diode is more or less fully conducting. It tries to pass all the current it can. To prevent the diode from destroying itself with excessive forward current, it will usually be used with a series resistance. Often this component will be specifically identified as a current-limiting resistor. Ohm's law defines the current flowing through the resistor by its resistance and the applied voltage. Since the diode is in series with the current-limiting resistor, the current flowing through it must be the same value. Therefore, the amount of current flowing through the diode is controlled by the external series resistance and not by the diode itself.

The most common application for a diode is rectification. In fact, diodes are often called rectifiers. Rectification converts an ac signal into pulsating dc. That is, one-half of each cycle is stripped off, so the output signal always has the same polarity. This is illustrated in Fig. 5-15.

If this pulsating dc output is filtered, as shown in Fig. 5-16, the original ac voltage will be converted into a dc voltage. This particular type of rectification is called half-wave rectification because only half of the ac waveform at the input is used. Half of the input energy is simply wasted and dissipated as heat.

A more efficient ac-to-dc converter can be designed using full-wave rectification. A very simple full-wave rectification circuit is illustrated in Fig. 5-17. This type of rectification requires

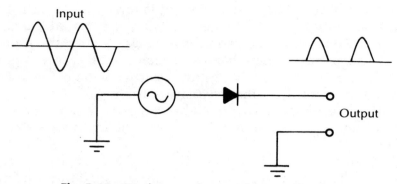

Fig. 5-15 *Diodes are often used for rectification.*

Fig. 5-16 *Filtering the pulsating output of a diode rectifier gives a closer approximation of a dc voltage.*

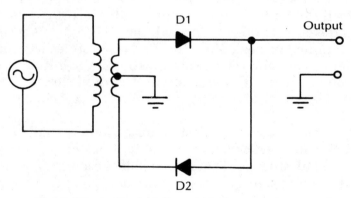

Fig. 5-17 *A simple full-wave rectification circuit.*

two diodes and a center-tapped transformer or similar ac voltage source. By grounding the center tap, the voltages seen by the two diodes are always 180 degrees out of phase. That is, they are of opposite polarities. During one-half of each input cycle, diode D1 is forward biased and diode D2 is reverse biased. During the other half of each input cycle, D1 is reverse biased and D2 is forward biased. That is, when D1 is passing a positive half-cycle, D2 is blocking a negative half-cycle, and vice-versa. One and only one of the diodes is conducting at a time (except when the input signal is actually crossing through the zero line). Both half-cycles are turned into dc pulses of the same polarity, as shown in Fig. 5-18. Besides the fact that less power is wasted, a full-wave rectification circuit's output is also easier to filter because there aren't large gaps between pulses.

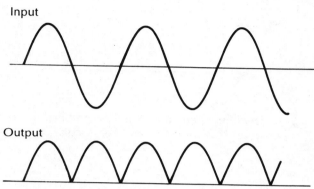

Fig. 5-18 *In the circuit of Fig. 5-17, both half-cycles are converted into dc pulses of the same polarity.*

A third type of rectification circuit is the bridge rectifier circuit, which is shown in Fig. 5-19. A bridge rectifier combines the major advantages of both full-wave rectifiers and half-wave rectifiers. Like the full-wave rectifier, the bridge rectifier uses the entire input cycle and is relatively easy to filter from pulsating dc to more or less true dc. Like the half-wave rectifier, a bridge rectifier does not require an expensive center-tapped transformer. Any two-ended ac voltage source can be used to drive a bridge rectifier circuit.

A bridge rectifier is composed of four diodes arranged in a square shape. Two (and only two) of these diodes are conducting at any instant. During positive half-cycles diodes D2 and D3 are forward biased, while D1 and D4 are reversed biased. This is reversed during negative half-cycles when D2 and D3 are reversed biased, while D1 and D4 are forward biased.

Bridge rectifiers are so useful, they are commonly marked as single monolithic components. A single component bridge rectifier has four leads corresponding to the connection points joining the four diodes contained within the single housing. A typical commercial bridge rectifier is illustrated in Fig. 5-20. Electrically, there is no difference between an integrated bridge

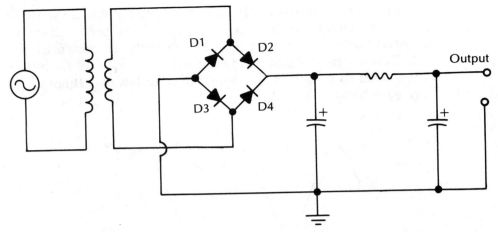

Fig. 5-19 *A bridge rectifier circuit combines the best features of the half-wave rectifier circuit and the full-wave rectifier circuit.*

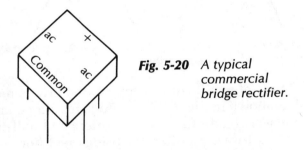

Fig. 5-20 *A typical commercial bridge rectifier.*

rectifier and four individual diodes arranged in a bridge rectifier pattern.

Diodes are also used in other applications, such as polarity protection. For example, a diode can prevent damage to a delicate circuit by blocking current flow (being reverse biased) when batteries are inserted backwards. Other common applications for the diode include switching circuits and wave shaping.

The most important specification for a semiconductor diode is the PIV (peak inverse voltage). Sometimes this specification is referred to as PRV (peak reverse voltage). There is no difference in meaning between these terms. In either case, the name is pretty much self-explanatory. The PIV rating is the maximum voltage that can be applied to a diode with a reverse bias without the diode breaking down. Once the diode breaks down (by exceeding the PIV), it will conduct current even though it is reverse biased.

With ordinary diodes, the PIV rating should never be exceeded. The diode could be damaged or destroyed if the PIV rating is exceeded for too long. With some diodes, "too long" could be less than 1 second. Other diodes are more durable.

Semiconductor diodes are also commonly rated for the maximum forward-biased voltage that can safely be applied. This is usually significantly higher than the PIV rating. It is never lower.

Another important specification for a semiconductor diode is its maximum power dissipation (how much power it can safely pass). This may be given in watts, or it may be listed as current rating. In either case, this rating must never be exceeded, even briefly, or it is likely that the diode will be damaged or destroyed.

For some applications, you may need to apply a larger reverse voltage than the largest available diode can safely handle. In such a case, you can effectively increase the PIV by using multiple diodes in series. To swamp internal differences in the individual diodes, a resistor and a capacitor are placed in parallel with each diode in the string, as illustrated in Fig. 5-21. All of these parallel capacitors should have equal values. The parallel resistors should also be equal.

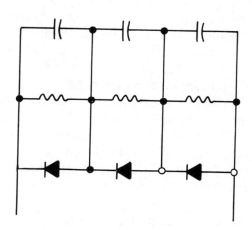

Fig. 5-21 *Internal differences in individual diodes in a string can be swamped with parallel resistors and capacitors.*

The capacitors ensure that any transients in the applied voltage will be shared equally by all of the component diodes in the string. The capacitance value used should be fairly small. Typically, it will be about three times the internal capacitance exhibited by the diode when it is zero biased (the applied voltage is 0 V). The exact capacitor value isn't terribly crucial, as long as it is close.

The resistor value is also selected in accordance with the internal characteristics of the diode it protects. The parallel resistance value should be approximately one-half the reverse resistance of the diode. The equation for the parallel resistance is

$$R = \frac{500\ V_b}{I_{r(max)}}$$

Where

V_b = peak voltage that will appear across each diode in the string, and

$I_{r(max)}$ = maximum specified reverse diode current, in milliamperes.

If you need to round off the resistance value given by this equation, always round down. The actual resistor value should not exceed the value given by the equation. In other words, the actual parallel resistance must be less than or equal to the calculated value.

Multiple diodes can also be connected in parallel to increase their effective current-handling capability. For the best results, the forward-biased resistances of the diodes in parallel should be closely matched. This will ensure that the applied current is divided equally between the paralleled diodes. If the ac resistances of the diodes are not identical, one of the diodes will conduct more than its partners.

To be on the safe side, it is a good idea to lower the maximum current-handling capability of the diodes by about 20%. This will leave you some headroom for unexpected transients or minor errors in the circuit calculations and component tolerances.

Practical diodes, like all other electronic components in the real world, are not perfect. If forward biased by a very low voltage, the diode might not conduct due to internal losses within the component. This is referred to as the diode's forward voltage drop. If the diode is used in a rectification circuit, a small portion of the passed half-cycle will be lost due to this forward voltage drop effect.

Most semiconductor diodes are made of either germanium or silicon. A typical germanium diode has a forward voltage drop of about 0.1 V to 0.3 V. Silicon diodes have higher forward voltage drops—typically 0.6 V to 0.8 V. The effect of the forward voltage drop of a diode is illustrated in Fig. 5-22.

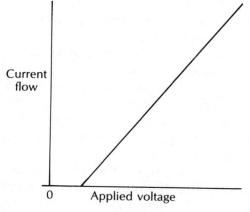

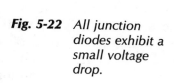

Fig. 5-22 *All junction diodes exhibit a small voltage drop.*

Zener diodes

There are several variants on the basic diode. One particularly common variation is the zener diode. The standard symbol used to represent zener diodes in schematic diagrams is shown in Fig. 5-23. Notice how this symbol differs from the symbol used to represent an ordinary diode. Again, the arrow in this symbol points toward the cathode, in the opposite direction of the forward-biased current (electron flow). The zener diode responds to a reverse polarity voltage in a rather unique way. To demonstrate this, we will use the simple circuit shown in Fig. 5-24.

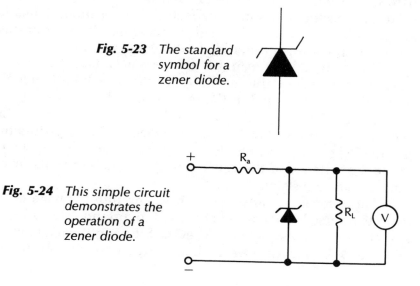

Fig. 5-23 *The standard symbol for a zener diode.*

Fig. 5-24 *This simple circuit demonstrates the operation of a zener diode.*

An important specification for a zener diode is the avalanche point, or breakdown voltage. The meaning of these names will be explained shortly. Unlike the case with an ordinary semiconductor diode, exceeding the breakdown voltage of a zener diode will do no damage to the component. In fact, normal applications for the zener diode always assume that the breakdown voltage will be exceeded. The avalanche point or breakdown voltage of a zener diode is usually fairly low. We will assume that the device we are using in our demonstration circuit (Fig. 5-24) is rated for 6.8 V (a typical value).

In this circuit, the zener diode is reverse biased. We will assume that the input voltage applied to this circuit is manually variable via a potentiometer. This variable voltage passes through a current-limiting resistance (R_a), before reaching the zener diode, which is in parallel with the load resistance (R_L) and a voltmeter indicating the voltage across the zener diode. For the purposes of our discussion, we will ignore the internal resistances of the zener diode and the voltmeter. Since the load resistance (R_L) is a constant value, altering the resistance of R_a will control the voltage seen by the cathode of the zener diode.

Let's start with the input voltage at 0 V. When the input voltage is increased to 1 V, the voltmeter reads just under 1 V. The small voltage drop is because of the resistance of current-limiting resistor R_a. This resistor protects the zener diode from the possibility of excessive current flow. The important thing to notice here is that with a 1-V (more or less) reverse bias on the zener diode, it does not conduct. In effect, the zener diode looks like an open circuit. As far as the load is concerned, it's the same as if the zener diode wasn't included in the circuit at all.

This will hold true until the source voltage exceeds the breakdown voltage rating of the zener diode, which is 6.8 V in our example. At this point, the reverse-biased zener diode starts to conduct. This is called the avalanche point of the diode because the current flow through the device abruptly rises from practically zero to a very high value. It is limited only by the low internal resistance of the zener diode and the external current-limiting series resistance (R_a).

The zener diode effectively sinks any input voltage greater than its reverse-biased avalanche point to ground. The voltmeter will continue to read just 6.8 V, even if the input voltage is now raised to 7 V, 8 V, or even higher. In other words, the reverse-biased zener diode limits the voltage applied to the load (R_L) to 6.8 V.

This basic zener diode circuit also serves to regulate the voltage applied to the load. That is, the voltage to the load remains reasonably constant, regardless of the amount of current drawn by the load. To understand this, let's see what happens if the load resistance (R_L) is varied with the zener diode omitted from the circuit. This revised version of our circuit is illustrated in Fig. 5-25.

Fig. 5-25 *Without the zener diode, the load resistance can significantly load down the supply voltage.*

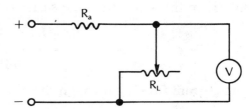

In the following example, we will assume that resistor R_a has a constant resistance of 500 Ω. We will also assume that the input voltage to this circuit is a constant 7 V. For simplicity, we will ignore the internal resistance of the voltmeter paralleled across the load resistance. In effect, R_a and R_L (the load resistance) form a simple voltage divider network. The voltmeter measures the voltage across R_L after the voltage drop across R_a has been subtracted.

First, let's say the load resistance (R_L) has a value of 100 Ω. Resistor R_a is in series with the load resistance (R_L), and resistances in series add, so the total resistance in the circuit is equal to

$$R_t = R_a + R_L$$
$$= 500 + 100$$
$$= 600 \ \Omega$$

Since we now know the total circuit resistance and the source voltage, we can use Ohm's law to find the amount of current flow through the circuit:

$$I = \frac{E}{R}$$
$$= \frac{7}{600}$$
$$= 0.0117 \ A$$
$$= 11.7 \ mA$$

Next, we can use this current value to find the voltage drop across the load resistance (R_L), which will be indicated on the voltmeter. (Remember, the same current flows through both resistances.)

$$E_L = IR_L$$
$$= 0.0117 \times 100$$
$$= 1.17 \text{ V}$$

Now, what happens if we increase the load resistance (R_L) to 800 Ω? In this case, the total series resistance is equal to

$$R_t = R_a + R_L$$
$$= 500 + 800$$
$$= 1300 \ \Omega$$

The current flowing through the circuit now works out to

$$I = \frac{E}{R_t}$$
$$= \frac{7}{1300}$$
$$= 0.00538 \text{ A}$$
$$= 5.38 \text{ mA}$$

This means that the voltage across the load resistance as indicated by the voltmeter will be about

$$E_L = IR_L$$
$$= 0.00538 \times 800$$
$$= 4.31 \text{ V}$$

Several other examples with various load resistance values are summarized in Table 5-1. Notice that the actual voltage across the load varies a great deal as the load resistance is changed. The load resistance has to be close to 10,000 Ω before the voltage seen by the load gets close to the desired 6.8 V (with a 7-V supply).

When the zener diode is in the circuit, however, it holds the output voltage to a fairly constant 6.8 V, provided the input voltage is higher than this. Since the load resistance is in parallel with the zener diode, E_L (the load voltage) is also a more or less constant 6.8 V. Remember, whenever any two components are in parallel, the voltage dropped across them is always equal.

**Table 5 – 1   Effects of varying the load resistance
in an unregulated circuit (*E* (source voltage) = 7 V, *R*ₐ = 500 Ω).**

R_L	R_t	*I* (mA)	Voltage across load (R_L)
100	600	11.7	1.17
200	700	10.0	2.00
300	800	8.8	2.63
400	900	7.8	3.11
500	1000	7.0	3.50
600	1100	6.4	3.82
700	1200	5.8	4.08
800	1300	5.4	4.31
900	1400	5.0	4.50
1000	1500	4.7	4.67
1100	1600	4.4	4.81
1200	1700	4.1	4.94
1300	1800	3.9	5.06
1400	1900	3.7	5.16
1500	2000	3.5	5.25
10,000	10,500	0.7	6.67

In short, what we have here is a simple voltage-regulation circuit. If the effective load resistance drops for any reason (that is, if the load circuit starts to draw more current), it will cause an increase in the voltage drop across resistor R_a (corresponding to the decreasing voltage drop across R_L). This decreases the voltage to the zener diode. As less voltage is applied to the diode, it draws less current. This means that the voltage drop across R_a must decrease, forcing the output (load) voltage to stabilize at the level determined by the avalanche point of the zener diode.

Of course, practical zener diodes can have breakdown voltages higher or lower than 6.8 V. This particular value was used here just as an example. Zener diodes are available with voltage ratings from less than 2 V up to about 200 V. Figure 5-26 is a graph illustrating the relationship of current and voltage through a typical zener diode.

The avalanche point of a zener diode is temperature sensitive. That is, it will change with variations in temperature. This is especially true for zener diodes with very low breakdown-voltage ratings (below about 6 V).

The zener diode is unique only when it is reverse biased. When forward biased, this component behaves pretty much like any ordinary diode.

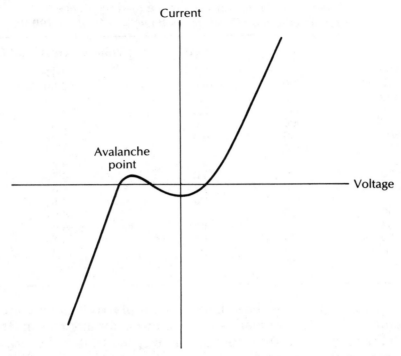

Fig. 5-26 *The relationship between the applied voltage and the current flow through a zener diode.*

Varactor diodes

Along with the semiconductor properties discussed in this chapter, the PN function of any semiconductor diode has a certain amount of internal capacitance. The exact amount of the internal capacitance in a diode depends primarily on the width of the junction itself. For most practical applications this internal capacitance is negligible and can be ignored. In some applications, it is undesirable. But in some specialized applications, the effect can be very useful, if it is controllable.

A varactor diode is a special type of diode designed to take advantage of this internal capacitance and its dependence on the junction width. Varying the reverse-biased voltage to a varactor diode will vary the effective size of the PN junction, which, in turn, alters the component's internal capacitance. In other words, the varactor diode functions as a voltage-variable capacitor. Because of the way it operates, a varactor diode is sometimes known as a voltage-controlled capacitor.

When a varactor diode is used as the capacitance in a resonant circuit (either series or parallel), the resonant frequency can be electrically controlled. Obviously, this device is ideal for automatic tuning systems and similar applications. Varactor diodes are also used in switching, limiting, harmonic generation, and parametric amplification circuits.

The symbol for a varactor diode is shown in Fig. 5-27. Notice the small capacitor symbol over the diode symbol. This small capacitor system is never shown as connected to anything in the circuit. It just represents the variable internal capacitance of the varactor diode. Also notice that while the diode symbol is usually filled in, the symbol for a varactor diode is often (though not always) shown in outline.

Fig. 5-27 *The symbol for a varactor diode.*

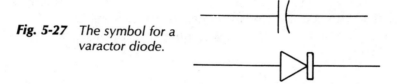

The capacitance of a varactor diode is usually relatively small. Generally speaking, 2000 pF is about the upper limit. Increasing the reverse-biased voltage decreases the capacitance. The Q (or quality factor) of a varactor diode is typically in the 50 to 300 range, with the upper end of this range being more common.

Snap-off diodes

Another type of diode that is designed with its internal capacitance in mind is the snap-off diode. Diodes are often used in electronic switching applications. When reverse biased, the diode appears to the circuit like an open switch, and when forward biased, it acts like a closed switch. A snap-off diode is specially designed to improve the switching characteristics of an ordinary junction diode.

The secret of a snap-off diode's operation is that when it is forward biased, a charge is stored in the diode. Once the diode has been charged, if it is then reverse biased, it will continue to conduct current until the stored charge has been used up. Ideally, the instant the charge is depleted, the diode should turn off completely. This ideal could only be achieved if the diode's internal capacitances and inductances were zero. This is not possible in a real-world component, but a snap-off diode is designed to keep these factors at an absolute minimum.

The snap-off diode is most commonly used in switching applications, but it may also be used in frequency multiplier circuits in the gigahertz (GHz) region. We are talking about extremely high frequencies here. One gigahertz equals 1000 MHz, or 1,000,000,000 Hz.

The input frequency is fed to the snap-off diode circuit. During the positive half of the input cycle, the diode is forward biased and a charge is stored in it. Then, when the input signal drops into its negative half-cycle, reverse biasing the snap-off diode, it conducts large amounts of reverse current then quickly reverts to the standard reverse-biased diode condition. The transient signal at the output of the snap-off diode is quite rich in high-order harmonics.

Hot-carrier diodes

Functionally related to the snap-off diode is the hot-carrier diode. Unlike the other semiconductor diodes discussed in this chapter, the hot-carrier diode does not have a PN junction. Instead, it features a junction between a semiconductor and a metal. The hot-carrier diode does not store a charge like the snap-off diode. There is virtually no overshoot when a hot-carrier diode is switched from a forward-biased condition to a reverse-biased condition, so its reverse recovery time is very fast. Because the hot-carrier diode can switch at extremely high frequencies, this component is often used in microwave circuits. Another advantage of the hot-carrier diode is that, unlike most other types of diodes, it generates very little noise. Also, the voltage drop across a forward-biased hot-carrier diode is very low, typically about 0.25 V.

PIN diodes

Another type of diode that is useful in high-frequency circuits is the PIN diode. This device can switch at rf frequencies of more than 300 MHz. Besides rf switching applications, PIN diodes are also often used in modulation circuits.

Most semiconductor diodes consist of two sections—each a differently doped slab of semiconductor material. A PIN diode, however, is comprised of three sections. The P-type section and the N-type section are separated by a layer of material with a very high resistance. This layer is called the intrinsic layer. The breakdown voltage of a PIN diode depends on the width of this intrinsic region.

There is a key frequency in the operation of a PIN diode. We will call this frequency F_0. As long as the applied signal frequency is below F_0, the PIN diode behaves pretty much like an ordinary junction diode. But if the applied signal frequency exceeds F_0, the PIN diode no longer rectifies. The diode action disappears and the device acts like a pure resistance. Yet, it is not a simple resistor, for the resistance is determined by the dc current flowing through the PIN diode. At room temperature, the high-frequency resistance is approximately equal to 48/I, where I is the dc current in milliamps. The high-frequency resistance of a typical PIN diode can vary from 1 Ω to 10 kΩ (10,000 Ω).

The specific F_0 frequency of a PIN diode is dependent on a specification called the recombination lifetime of the device. The recombination lifetime of a PIN diode is abbreviated as r. The formula for finding the F_0 frequency is:

$$F_0 = \frac{1}{2\pi r}$$

$$= \frac{1}{6.28r}$$

Shockley diodes

Another type of special-purpose diode is the Shockley diode. Unlike other types of semiconductor diodes, the Shockley diode has more than a single PN junction. Its construction includes two of each type of semiconductor in an alternating pattern. That is, where an ordinary diode can be described as NP, the Shockley diode can be described as NPNP. Because it is made up of four semiconductor sections, the Shockley diode is also known as a four-layer diode. Like most other diodes, the Shockley diode has two terminals—an anode and a cathode. This is why it fits the definition of diode, which means a two-electrode device. The symbol for a Shockley diode, shown in Fig. 5-28, is quite different from the standard diode symbol.

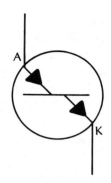

Fig. 5-28 *The symbol for a Shockley (or Schottky) diode.*

The Shockley diode is used primarily in switching applications. This component features switching properties similar to those of a neon glow lamp. Each Shockley diode has an inherent trigger voltage. Below the trigger voltage, the device is in its "off" state and it exhibits a very high resistance. But if the applied voltage exceeds the trigger value, the Shockley diode will be switched on and the resistance will drop to an extremely low value. Typically, the on resistance of a Shockley diode is just a few ohms. This is significantly lower than the forward-biased resistance of most ordinary diodes.

The trigger voltage may be identified by any of several alternate names in various technical manuals and specification sheets. The trigger voltage may be called threshold voltage, firing voltage, or avalanche voltage. All of these terms are interchangeable when talking about a Shockley diode.

If a third electrode, called a gate, is added to a four-layer Shockley diode, the result is a component known as a silicon controlled rectifier (SCR). This more complex device will be discussed in chapter 6.

Fast-recovery diodes

Most ordinary diodes are designed so that they give their best performance at relatively low frequencies. One of the most common applications for a diode is to rectify 60-Hz sine waves. This is the frequency of normal ac power lines in the United States.

A certain finite amount of time is required for a diode to recover. This is the time it takes for the diode to turn off when the polarity of the applied voltage is reversed. This normally takes just a fraction of a second. In low-frequency applications, the recovery time of the diode is not particularly significant. However, in high-frequency applications, such as in a television fly-back circuit, the diode recovery time can become very crucial. This is because the diode must respond to very short-duration spikes, with a very brief "rest" period between adjacent spikes. An ordinary diode could cause erratic or incorrect operation of the circuit. For better and more reliable performance in high-frequency circuits, a special-purpose diode called a fast-recovery diode is used.

LEDs

When current passes through any substance, some of the electrical energy is converted into heat energy. A semiconductor diode can be designed so that much of this energy is converted into light instead of heat. This type of device is called a light-emitting diode (LED).

Like an ordinary diode, an LED will pass current in only one direction. That is, it is a polarized device. When an LED is reverse biased, nothing special happens. It simply blocks the flow of current, just like an ordinary diode. But when the LED is forward biased, it glows, or emits light energy.

While some clear LEDs have been developed, most LEDs emit colored light. Red is by far the most common color for an LED, but green and yellow are also frequently used. Recently, blue LEDs have been developed, but they don't appear to be available on the hobbyist market yet. In addition to these visible color types, some LEDs are designed to emit light in the infrared region.

Figure 5-29 shows the two most commonly used symbols for LEDs. There is no meaningful difference between these two symbols. The circle in Fig. 5-29B doesn't add any information to the symbol. The important factor here is the two (sometimes three) small arrows pointing out away from the body of the component. This indicates the emission of light from the device.

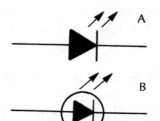

Fig. 5-29 *The symbols for an LED.*

LEDs are used in electronic circuits primarily as indicator devices. That is, an operator can easily tell (even at a distance) whether or not a specific voltage is present in a circuit by whether the LED is lit up or dark. For example, in the simple circuit shown in Fig. 5-30, the LED lights up whenever the circuit is activated (that is, the switch is closed). This could serve as a reminder of which subcircuits are currently active or to turn off the equipment when not in use. When this LED is lit, it also indicates that the circuit is getting power, and is presumably operating properly.

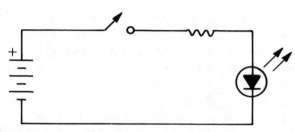

Fig. 5-30 *This simple circuit demonstrates the operation of an LED.*

Within certain limits, the higher the voltage applied to an LED, the brighter it will glow. Of course, lowering the applied voltage will dim the glow of the LED. This cannot be used to practically measure exact values, but it can sometimes be used for relative comparisons.

LEDs are relatively durable, but they can be damaged if they are misused. LEDs are intended for use in low-power circuits only. Typically, no more than 3 V to 6 V should be applied to an LED, especially for extended periods. A typical LED will glow with an applied voltage as low as 1.3 V. There is some leeway in the maximum applied voltage to an LED. Often LEDs are used in 9-V circuits, but to avoid premature failure, the component should be protected with a simple voltage divider network of some kind. Excessively high power levels could burn out the semiconductor junction and render the LED useless. Some consideration should also be given to the amount of current flowing through an LED. Excessive current can damage or destroy the component.

The LED, being a diode, exhibits a very high resistance when it is reverse biased. According to Ohm's law ($I = E/R$), this means that the diode will draw very little current under this condition. When forward biased, on the other hand, the LED's internal resistance drops to a very low value, allowing the current flow to climb to a relatively high level. Depending on the rest of the circuitry involved, the LED may attempt to pass more current than it can safely handle.

To limit the current flow through an LED to a safe value, a series resistor is almost always added to any LED circuit, as illustrated in Fig. 5-31. This resistor should have a relatively low value, usually between 100 and 1000 Ω. Some sources recommend a current-limiting series resistance no higher than about 600 Ω. The higher this current-limiting resistance is, the dimmer the glow of the LED will be. A lower current-limiting resistance will increase the brightness of the LED, but the current flow through the component will also be increased. A good average value for the current-limiting series resistor in most LED circuits is 330 Ω. This is a standard resistor value. The tolerance of a resistor in this application is usually not crucial, unless the resistance value is close to one of the acceptable extremes or if the application calls for reasonable matching of the brightness of multiple LEDs.

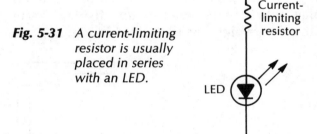

Fig. 5-31 *A current-limiting resistor is usually placed in series with an LED.*

Multiple-segment LED displays

Often multiple LEDs are used together to offer more detailed information than simple on-off indication. For example, several LEDs can be arranged in a row, as shown in Fig. 5-32, to form what is known as a bar graph. Usually the LEDs will be driven by various points in a multistage voltage divider network. The higher the input voltage is, the more LEDs will be lit.

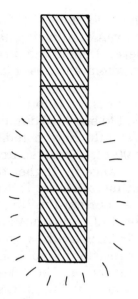

Fig. 5-32 *Several LEDs can be arranged in a row to create a bar graph.*

A similar display mode is the dot graph. In this case, only one LED representing the highest activated voltage point in the voltage network is lit. All higher and lower LEDs are dark.

Bar graphs and dot graphs can be built from individual LEDs, or manufactured multiple-display units in convenient single housings are available. Probably the most common multiple-LED display unit is the seven-segment display. Seven LEDs are arranged in a figure eight, as illustrated in Fig. 5-33. Lighting up

Fig. 5-33 *A seven-segment display has seven LED segments arranged in a figure eight.*

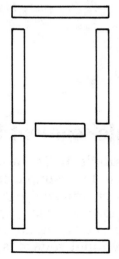

selected LED segments in the display, while leaving others dark, permits the display of any numeral from 0 to 9. Certain alphabetic letters can also be displayed. Of course, multiple seven-segment displays can be used together to display higher numerical values.

A seven-segment display unit has eight leads—one for each segment and one common lead for the opposite end of each segment. Some seven-segment LEDs have a common cathode, as shown in Fig. 5-34. Others have a common anode, as shown in Fig. 5-35.

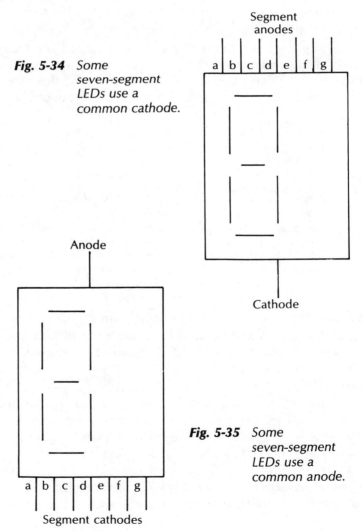

Fig. 5-34 *Some seven-segment LEDs use a common cathode.*

Fig. 5-35 *Some seven-segment LEDs use a common anode.*

Seven-segment LED display units are not as widely used as they were a few years ago. They are gradually being replaced by similar devices using LCDs (liquid-crystal displays). LCDs consume less power than LEDs, and they are easier to read under bright lighting conditions. LCDs also tend to be more versatile than LEDs. LCDs will be discussed in chapter 9.

Tristate LEDs

Since LEDs will glow only if they are forward biased, they are ideal for checking the polarity of a voltage. Figure 5-36 shows the circuit for a simple but effective polarity checker. The single resistor limits the current through both of the LEDs. Only one current-limiting resistor is needed, because only one of the LEDs will ever be lit at any given instant. Obviously, both opposite polarity diodes cannot be forward biased simultaneously.

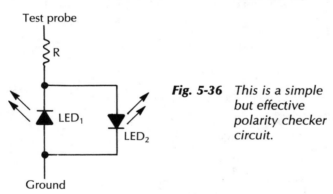

Fig. 5-36 *This is a simple but effective polarity checker circuit.*

For the most convenient and least ambiguous indication of the signal polarity, the two LEDs in this circuit should be of contrasting colors. For example, LED$_1$ could be red, while LED$_2$ could be green. If the test lead is touched to the voltage source, we can easily tell the polarity of the unknown signal. If the voltage is positive with respect to the ground, red LED$_1$ will light up, and green LED$_2$ will remain dark. If the polarity is reversed so that the tested voltage is negative with respect to ground, green LED$_2$ will light up, while red LED$_1$ remains dark.

If the two LEDs alternately blink on and off, the tested voltage signal is ac rather than dc. If the signal frequency is moderate to high (above 10 to 15 Hz), the human eye will not be able to catch the individual blinks, so both LEDs will appear to be continuously lit, though possibly at less than normal brightness.

This simple circuit is so useful that semiconductor manufacturers offer it in a single package. The current-limiting series resistor may be built into the device, or it may need to be externally added.

Such a two-in-one LED is known as a tristate LED, or a three-state LED, because it has three possible conditions (excluding off—both LEDs dark). With one polarity the red LED will light up and the green LED will be dark. With the opposite polarity, the green LED will be the one glowing, while the red LED is dark. If the input voltage is an ac signal, the two LEDs will blink on and off or both LEDs will appear to be continuously lit. The two LEDs are very close together and under a single diffusing lens, so their colors (red and green) will blend together and the tristate LED will appear to be glowing yellow.

We can summarize the action of a tristate LED as follows:

- Dark—no input voltage;
- Red—dc input voltage with polarity A;
- Green—dc input voltage with polarity B; and
- Yellow—ac input voltage.

Clearly, the tristate LED is a very useful and versatile indicating device.

The standard symbol for a tristate LED is shown in Fig. 5-37. In this case, the circle is usually employed to indicate that this is a single device rather than two separate LEDs.

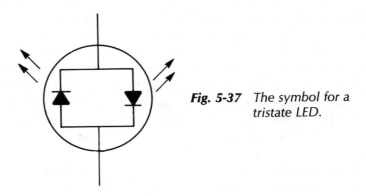

Fig. 5-37 *The symbol for a tristate LED.*

Laser diodes

Another very special type of diode is the laser diode. You can almost consider a laser diode to be a super souped-up LED. It emits light, but a much more powerful and more focused type of light. The word *laser* is actually an acronym meaning light amplification by stimulated emission of radiation.

Most light is multichromatic, incoherent, and diffused over a wide area. Laser light is an exception to this. Multichromatic means light of more than one color is included. Different colors have different frequencies. Most natural light sources generate several different frequencies. White light, of course, consists of almost equal amounts of all possible color frequencies. But even a red light might include multiple shades of red, and even a few completely nonred frequencies, such as blue or green. A laser light, on the other hand, is monochromatic. It is very pure, consisting of just a single color frequency. Laser light can be considered the light equivalent of the sine wave.

When we say a light source is incoherent, we mean that the various light waves making up the total have random phase relationships. Most of the light waves are out of phase with one another. This means they can strengthen each other at some points, but oppose and weaken each other at other points. A laser light is coherent. All of its light waves are in phase with one another, increasing their effective power. A highly coherent laser beam of just a few watts can actually burn through most materials.

Finally, most ordinary light sources are quite diffuse. The light waves spread out from the source in all directions. This is illustrated in Fig. 5-38. A common flashlight emits a beam of light through a lens of just an inch or two in diameter. But if we aim the flashlight at an object a few feet away, we find that the light beam has spread out to a diameter of several feet. Obviously, this light beam cannot travel very far before it becomes so spread out and diffuse that it is too weak to detect. A laser beam, however, is very tightly focused and tends to stay that way. If we aim a high-quality laser beam with a diameter smaller than your little finger at the moon, by the time it gets to the lunar surface, it will have diffused only enough to have a diameter slightly larger than a few yards.

In other words, an ordinary light source wastes a lot of energy. Its wave components fight each other, and the available energy is diffused over a wide area. A laser light source, however,

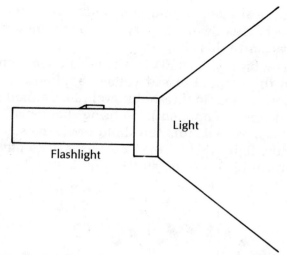

Fig. 5-38   *The light waves from an ordinary source quickly spread out in all directions.*

wastes very little energy. Its wave components all work together and are tightly packed, so all the energy is concentrated into a very small area.

Early lasers were quite large and expensive, usually requiring costly, bulky crystals and delicate glass tubes. The relatively recent development of the semiconductor laser diode has brought both the size and cost of lasers down considerably.

A typical laser diode looks something like the drawing in Fig. 5-39. Notice that there is only one lead. The other electrical connection is made directly to the threaded base of the component.

Fig. 5-39   *A typical laser diode.*

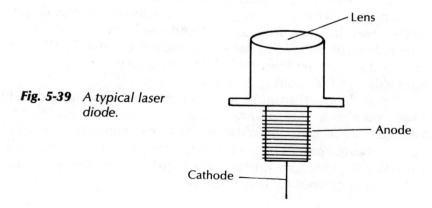

The technical name of the semiconductor laser diode is the injection laser. This device is very similar to the common LED discussed earlier in this chapter.

To review briefly, an LED is a two-lead semiconductor device. When the forward-biased voltage applied across the two leads is below a specific threshold level (determined by the construction of the component), nothing happens. When the applied voltage exceeds this threshold level, the semiconductor junction emits light; that is, the LED glows. The physical structure of a typical LED is illustrated in Fig. 5-40.

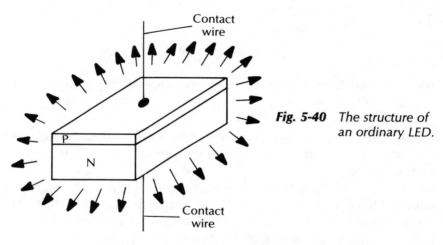

Fig. 5-40   *The structure of an ordinary LED.*

At low voltages a laser diode works in pretty much the same way. But an injection laser diode also has a second critical threshold point called Jth, or Ith. These two terms are interchangeable.

Below the Jth threshold, the diode functions exactly like an ordinary LED. The junction glows with a relatively broad spectrum of wavelengths (frequencies) and the light is emitted in a wide pattern of radiation (that is, the light is emitted in all directions, not in a single beam). Under these conditions, the diode is obviously not functioning as a laser device.

But if the Jth threshold level is exceeded, the emitted light thins down to a very narrow beam. The emitted light is now coherent and of a single frequency (monochromatic). The emitted laser beam escapes from both of the laser end faces, unless, as is usually the case, one end is coated with a reflective film. Gold is usually employed for this purpose.

A simplified diagram of the internal construction of a typical injection laser diode is illustrated in Fig. 5-41. Compare this with the structure of an ordinary LED, shown in Fig. 5-40. The operational differences between an ordinary LED and an injection laser diode are shown in Fig. 5-42.

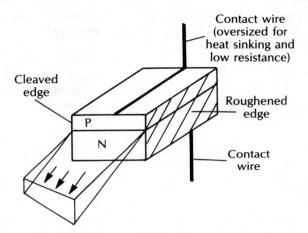

Fig. 5-41   *The structure of an injection laser diode.*

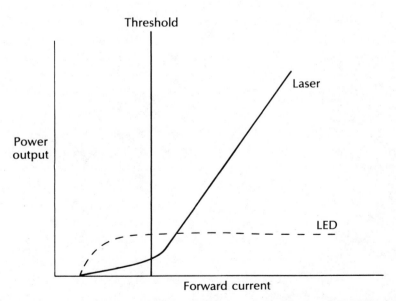

Fig. 5-42   *The operational differences between an ordinary LED and a laser diode.*

A laser diode will only work when it is forward biased. Most laser diodes can withstand only a relatively small reverse-biased voltage. If too high a reverse-biased voltage is applied to a laser diode, the component may be damaged or destroyed.

The standard symbol for a laser diode is shown in Fig. 5-43. Notice how the light emission arrows are zigzagged, rather than straight as in an ordinary LED symbol. This is done to indicate that the symbol is not referring to ordinary light.

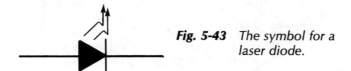

Fig. 5-43   *The symbol for a laser diode.*

Transistors

A SEMICONDUCTOR DIODE IS CERTAINLY USEFUL IN ELECTRONIC circuits, but it is still a fairly passive device. It does not amplify the signal. The semiconductor component that really revolutionized the world of electronics is the transistor. Where a diode has just a single *PN* junction and two leads, the transistor (usually) has two *PN* junctions and three leads. Transistors do exhibit gain and are true active devices. The name "transistor" comes from transfer resistor.

Bipolar transistors

The simplest and most common type of transistor is the bipolar transistor. This type of transistor is like a pair of merged diodes, with two *PN* junctions. There are two types of bipolar transistors. They are called *NPN* transistors and *PNP* transistors. Both can be described as semiconductor sandwiches.

In an *NPN* transistor, as shown in Fig. 6-1, there is a very thin slice of *P*-type semiconductor sandwiched between two thicker slabs of *N*-type semiconductor. Leads extend from each of the semiconductor sections. Obviously, the name is just a description of the type and arrangement of the semiconductor sections in the device. We have an *N*-type section, then a *P*-type section, then another *N*-type section, so it is an *NPN* transistor.

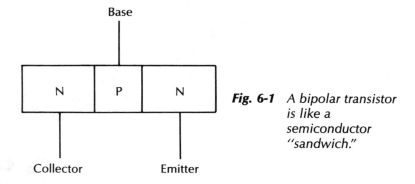

Fig. 6-1 *A bipolar transistor is like a semiconductor "sandwich."*

You've probably already guessed that a *PNP* transistor just reverses the arrangement of the semiconductor types, as illustrated in Fig. 6-2. Here we have a thin slice of N-type semiconductor sandwiched between two thicker slabs of P-type semiconductor.

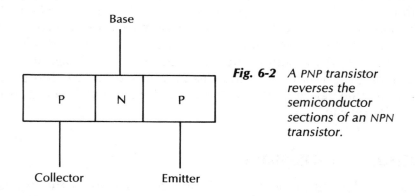

Fig. 6-2 *A PNP transistor reverses the semiconductor sections of an NPN transistor.*

In the following discussion, we will concentrate on the NPN transistor, for convenience. The *PNP* transistor works in exactly the same way as the *NPN* transistor, except all polarities are reversed.

One of the end sections in an NPN transistor is identified as the emitter. The other N-type section is the collector. The P-type section in the middle is called the base. Each of these terms will be explained shortly.

The entire semiconductor sandwich is enclosed in a protective plastic or metal case. When the case is metal, one of the leads is often (though not always) electrically connected to the case. Most frequently, this is the collector, but there are some exceptions. When in doubt, check the manufacturer's data sheet, or use an ohmmeter to test for continuity (zero resistance) between each of the leads and the transistor's case. Of course, this is irrelevant if the transistor is enclosed in a plastic housing.

Low-power transistors usually look something like Fig. 6-3. For higher power applications, a power transistor is used. This type of transistor looks like one of the drawings in Fig. 6-4.

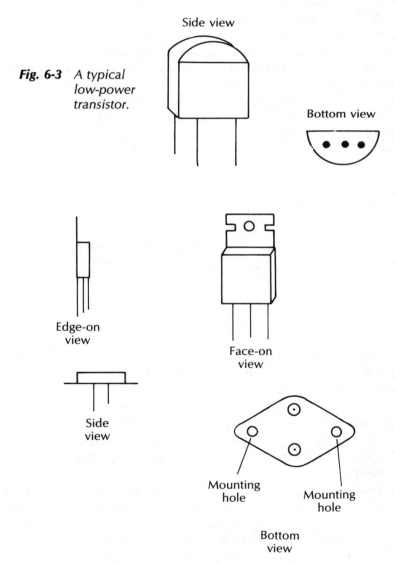

Fig. 6-3 *A typical low-power transistor.*

Side view

Bottom view

Edge-on view

Face-on view

Side view

Mounting hole

Mounting hole

Bottom view

Fig. 6-4 *Some typical power transistors.*

Notice that one of these power transistors has only two leads. The third connection (almost always the collector) is made directly to the metallic case of the transistor. Notice also that both types of power transistors have relatively large metal areas that can be screwed directly to a chassis. This large metal area serves as a heat sink. A heat sink's purpose is to conduct heat away from the delicate semiconductor crystal and dissipate it into the surrounding air. If a transistor is used at power levels near its rated limit, an additional external heat sink is strongly advised. Too much heat sinking will never affect the operation of the circuit, but if a transistor is allowed to run too hot, it can easily self-destruct.

The symbol used to represent an *NPN* bipolar transistor is shown in Fig. 6-5. Many technicians omit the surrounding circle because it doesn't add any extra information to the symbol. Other technicians feel the circle makes the schematic diagram look a little neater, and makes the transistor symbol easier to see at a glance.

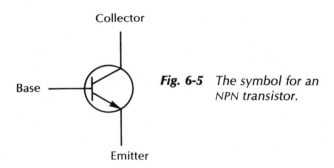

Fig. 6-5 *The symbol for an NPN transistor.*

Notice that three separate and independent leads are shown. The lead marked E is the emitter, B is the base, and C is the collector. Some schematics have the leads marked in this manner, but usually it is assumed that you can tell from the symbol which lead is which. The lead with the arrow is always the emitter. The emitter and the collector are usually doped somewhat differently in a practical transistor, so they are rarely electrically interchangeable.

The symbol for a *PNP* bipolar transistor is shown in Fig. 6-6. Notice that the only difference here is the direction the arrow in the emitter is pointing. A handy way to remember which is which is *NPN* never points in, and *PNP* points in perpetually.

Unfortunately, there isn't much standardization of the lead positions on actual transistors. Some possible arrangements are

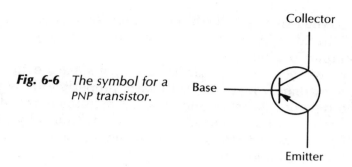

Fig. 6-6 *The symbol for a
PNP transistor.*

shown in Fig. 6-7. There is no way to tell which lead is which
just by looking at the transistor. For this reason, every electronics
technician and hobbyist should have a good transistor specification book that identifies the leads on various transistors. You can
also go by the manufacturer's specification sheet, if it is available. Transistors are often sold without the specification sheet. It
can usually be requested from the manufacturer, although there
is often an extra charge for this.

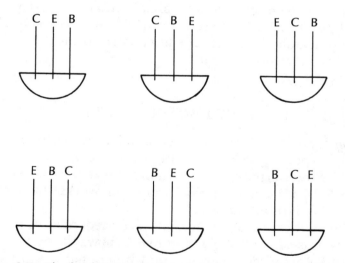

Fig. 6-7 *The lead placement on transistors is not standardized.*

Most transistor specification books are also substitution
guides. Literally thousands of different transistors have been
manufactured all over the world over the years, and many specific types are difficult, if not impossible, to locate. Fortunately,
many transistor types are interchangeable (at least in most cir-

cuits), so you can often substitute one type for another. A transistor substitution guide cross-references transistor type numbers with similar specifications.

However, you should bear in mind that some electronic circuits are extremely fussy and will work properly only with one specific type of transistor. Such special requirements will usually be noted on the schematic or parts list. Generally there is considerable leeway in the substitution of transistors. This is particularly useful for electronics hobbyists who usually don't have the wherewithal to locate a source for every possible transistor type number. A good transistor substitution guide is really an absolute necessity for anyone working in electronics, whether professionally or as a hobby.

Some projects may not list a transistor type number at all. This is usually because the exact specifications of the transistor don't matter very much in the particular application. Almost any general purpose transistor can be used in such a circuit, provided it can handle the necessary power and current levels. For general purpose *NPN* transistors, the 2N2222 or the 2N3904 are usually good choices. They are widely available and inexpensive. A good general purpose *PNP* transistor is the 2N3906. The *PNP* 2N3906 is closely matched (similar specifications) to the *NPN* 2N3904, except for the reversed polarities, of course.

How a bipolar transistor works

In order for a bipolar transistor to function properly in a circuit, the correct polarity relationship must be set up among its three leads. For the purposes of discussion, we will be concentrating on the *NPN* transistor. A *PNP* transistor works the same way, except all polarities are reversed.

The application of voltages of the correct relative polarity to a transistor is called biasing. Figure 6-8 shows the correct biasing for an *NPN* transistor. This is not a functional circuit. It is for illustrative purposes only.

In most applications, the collector of a correctly biased *NPN* transistor is positive with respect to the base, and the emitter is negative with respect to the base. In Fig. 6-8 this is accomplished with two separate voltage sources for maximum clarity. In most practical circuits, some sort of voltage divider network is used to derive the correct relative polarities for biasing.

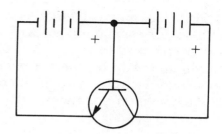

Fig. 6-8 *The correct biasing of an NPN transistor.*

Notice that we are talking about relative polarities here, not absolute polarities. The polarities may or may not be correct with respect to ground. The collector must have the highest (most positive or least negative) voltage, and the emitter must have the lowest (least positive or most negative) voltage. The base voltage is somewhere in between the collector voltage and the emitter voltage. All three voltages may be positive with respect to ground, or all three may be negative with respect to ground, or they may be a combination of positive and negative voltages. This doesn't matter, as long as the correct relationship is maintained between the collector, base, and emitter voltages.

You should recall that an N-type semiconductor has extra electrons and a P-type semiconductor has extra holes (spaces for missing electrons). Since the P-type semiconductor section in an NPN transistor is much thinner than either of the N-type sections, it logically has fewer holes than the end sections have spare electrons. The negative charge from the terminal connected to the emitter forces the spare electrons in the N-type emitter section towards the P-type base region. The base-emitter PN junction is forward biased, so these electrons can cross over into the base section, filling the free holes. But there are too many loose electrons and not enough free holes.

Because the base section now has more electrons than in its normal state, it acquires an overall negative charge that forces the extra electrons out of the base region. Of course, these electrons cannot cross back over the base-emitter PN junction because of the strong negative charge being applied to the emitter section. Some electrons will leave the transistor through the base lead to the positive terminal of the base-emitter battery because the base lead is kept positive with respect to the emitter. But this base current flow is small.

The collector lead is at an even more positive potential. The excess electrons are drawn out of the N-type collector section into the positive terminal of the collector-base battery. This leaves this section of the semiconductor with a strong positive charge. This positive electrical charge on the collector pulls most of the extra electrons out of the base region across the base-collector *PN* junction. These electrons jump the junction into the collector section and are drawn off into the positive terminal of the external voltage source.

We can simplify all of this by saying that the transistor's leads are accurately named. The emitter emits electrons and the collector collects them.

About 95% of the transistor's total current flow will pass through the collector, while only about 5% of the total current will leave the transistor through the base lead. Just how much current is drawn into the transistor by the emitter is determined by the individual characteristics of the individual transistor being used, and the actual voltage being applied to the component's base terminal.

We can adapt the basic circuit of Fig. 6-8 to permit a manually variable base voltage by adding a potentiometer, as shown in Fig. 6-9. Adjusting this potentiometer will control the voltage to the base, which will, in turn, determine the amount of current flow through the transistor. Regardless of the total amount of current drawn, only about 5% will flow through the base lead, and the remaining 95% will flow out of the collector lead through a load resistance.

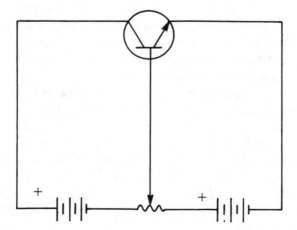

Fig. 6-9 *Adjusting the potentiometer in this circuit applies a varying voltage to the base of the transistor.*

A very small change in the transistor's base current will result in a very large change in the collector current. For this reason, transistors are sometimes called current amplifiers. A transistor amplifies current rather than voltage or power. Of course, thanks to Ohm's law, the net effect is basically the same because varying the current flow through the load resistance will proportionately vary the voltage dropped across it.

Most practical transistor circuits use just a single voltage source, and the intermediate base voltage is derived via a voltage divider network, as shown in the circuit of Fig. 6-10. In this circuit, the voltage divider network is comprised of resistors R_1, R_2, and R_3. Note that if a very small ac signal voltage is also applied to the base of the transistor, the voltage on the base will vary above and below its nominal dc value. This is comparable to adjusting the potentiometer in the circuit of Fig. 6-9. This changing base voltage causes the collector current and, thus, the voltage drop across the load resistance (R_L) to vary in step with the varying input voltage being fed into the base. Because of the current gain of the transistor, the ac voltage across the load (R_L) will be much larger (higher amplitude) than the original ac source voltage at the input (base). In other words, the ac input signal is amplified by the transistor.

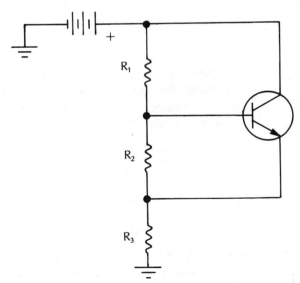

Fig. 6-10 *A single supply voltage can be used to bias a transistor by employing a voltage divider network.*

Alpha and beta

There are literally thousands of different types of bipolar transistors available. They differ in a number of factors, or specifications. For example, two transistors might differ in the maximum amount of power they can safely dissipate, their internal impedances, and, most importantly, how much current gain they can produce. The two most important specifications for a transistor are a pair of interrelated gain factors called alpha and beta. Alpha is indicated by the Greek letter α. The symbol for beta is β.

Alpha is the transistor's current gain between the emitter and the collector. That is, for any given change in the emitter current (with the supply voltage held constant), the collector current will change with a fixed relationship (α) to the emitter current. The basic equation for determining the alpha of a bipolar transistor is

$$\alpha = \frac{\Delta I_C}{\Delta I_E}$$

The small triangular symbol (Δ) is read as delta. It is used to identify a changing value. I_C is the collector current, and I_E is the emitter current. Alpha can also be defined in terms of dc. In this case, the delta symbols are omitted from the formula:

$$\alpha = \frac{I_C}{I_E}$$

As a typical example, let's suppose we have a transistor in which a 2.6-mA (0.0026-A) change in the emitter current results in a 2.4-mA (0.0024-A) change in the collector current. The alpha of this hypothetical transistor works out to

$$\alpha = \frac{\Delta I_C}{\Delta I_E}$$

$$= \frac{2.4}{2.6}$$

$$= 0.92$$

Notice that the collector current always changes less than the emitter current does. This is because there is always a negative current gain from the emitter to the collector in a bipolar transistor. Remember, only about 95% of the emitter current gets through to the collector. For any practical bipolar transistor, alpha will always be less than unity (one).

A small change in the base current, however, results in a large change in the collector current. Only about 5% of the total current through a transistor flows through the base lead, while the remaining 95% flows through the collector lead. For our sample transistor, the same 2.4-mA (0.0024-A) current change in the collector can be achieved with a mere 0.2-mA (0.0002-A) change in the base current. The ratio between the base current and the collector current is β. The formula for beta is

$$\beta \; = \; \frac{\Delta I_C}{\Delta I_B}$$

I_C is the collector current, and I_B is the base current. Once again, the triangular delta symbol represents a changing value. The dc form of the beta equation is simply

$$\beta \; = \; \frac{I_C}{I_B}$$

For our sample transistor, the beta value works out to

$$\beta \; = \; \frac{\Delta I_C}{\Delta I_B}$$
$$= \; \frac{2.4}{0.2}$$
$$= \; 12$$

Beta is always greater than one, indicating a positive current gain from the base to the collector.

The alpha and beta values for any given transistor are closely interrelated. If you know one, you can determine the other. For example, if you know the alpha value and need to find the corresponding beta value, you can use the equation

$$\beta \; = \; \frac{\alpha}{(1 \; - \; \alpha)}$$

Similarly, if you know the beta value, the formula for finding the alpha value is

$$\alpha = \frac{\beta}{(1 + \beta)}$$

For example, the beta for our sample transistor was approximately 12, so we can calculate the alpha as

$$\alpha = \frac{\beta}{(1 + \beta)}$$

$$= \frac{12}{(1 + 12)}$$

$$= \frac{12}{13}$$

$$= 0.92$$

Of course, this is the same value we found when we figured the alpha directly from the emitter and collector currents.

As a second example, let's assume we have a transistor with an α of 0.88. The β of this transistor would be equal to

$$\beta = \frac{\alpha}{(1 - \alpha)}$$

$$= \frac{0.88}{(1 - 0.88)}$$

$$= \frac{0.88}{0.12}$$

$$= 7$$

For our last example, we will assume the transistor has an α of 0.97. This time the beta value works out to

$$\beta = \frac{\alpha}{(1 - \alpha)}$$

$$= \frac{0.97}{(1 - 0.97)}$$

$$= \frac{0.97}{0.03}$$

$$= 32$$

As you can see, alpha and beta always increase together. As one gets larger, so does the other.

Other transistor specifications

Alpha and beta are very important, of course, but they aren't the only important specifications for a bipolar transistor. We will briefly discuss just a few of the more important transistor specifications here.

The total current flowing through the collector should be equal to alpha multiplied by the emitter current. That is,

$$I_C = \alpha \, I_E$$

When working with a practical transistor, however, this simple equation isn't too accurate. There is another factor contributing to the total collector current. This is the leakage current flowing through the reverse-biased base-collector junction. The leakage current is identified as I_{CBO}, or sometimes just I_{CO}. This is the current that would pass from the collector to the base junction if the emitter was left open in a common-base configuration.

A similar parameter is the I_{CEO}, or as it is occasionally written, I_{EO}. This is the collector-to-emitter current that would flow through the transistor if the base lead was left open (in a common-emitter configuration).

The manufacturer's spec sheet for a transistor will also include several voltage specifications that define the maximum operating limits of the device in question. The voltage used to forward bias the emitter-base junction is called V_{EE}, and the reverse-bias voltage for the collector-base junction is known as V_{CC}. The manufacturer will specify maximum values for each of these supply voltages. Excess voltage to a transistor can result in either an avalanche or punch-through breakdown. The delicate semiconductor crystal within the transistor could be damaged or destroyed. The specification for the collector-base avalanche breakdown voltage is identified on most manufacturer's spec sheets as BV_{CBO}.

If the base of a transistor is left open and the emitter is connected to the input voltage supply instead of the base, there can be a breakdown between the collector and the emitter. In this case, the breakdown voltage is called BV_{CEO}. Typically, the BV_{CEO} specification is about one-half the BV_{CBO} value. The BV_{CEO} voltage limit can be increased by connecting external resistors or other dc load resistances between the base and the emitter, rather than leaving the base floating. The collector-to-emitter break-down voltage increases as the dc resistance between the base and the emitter is decreased.

Transistor amplifier configurations

There are three basic ways a bipolar transistor can be used as an amplifier. Each is identified by which of the transistor's three leads is common to both the input and output circuits. Not surprisingly, these three transistor amplifier configurations are called common-emitter, common-base, and common-collector. Each of these transistor amplifier circuits has different characteristics and each is suitable for different applications.

The common-emitter amplifier

The basic common-emitter amplifier circuit is shown in Fig. 6-11. Notice that the transistor's emitter lead is used here as the common reference point for both the input signal and the output signal. As you can see, the input signal is fed into the transistor's base and the output signal is tapped off from the collector lead.

While it isn't always necessary, most practical common-emitter amplifier circuits include a resistor between the emitter

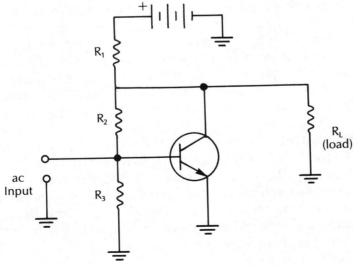

Fig. 6-11   The basic common-emitter amplifier circuit.

of the transistor and the actual ground point of the circuit. This resistor is used to improve the stability of the circuit. The emitter is still considered to be grounded and for all intents and purposes is effectively at ground potential 0 V.

The common-emitter configuration exhibits a low input impedance, minimizing loading effects on any preceding stages. Most typical common-emitter amplifier circuits will have an input impedance somewhere between 200 and 1000 Ω (1 kΩ). The common-emitter amplifier circuit also features a high output impedance, so it is relatively unaffected by loading effects from any later stages. Typical output impedances for a common-emitter amplifier circuit range from about 10 kΩ (10,000 Ω) to 100 kΩ (100,000 Ω). Power can be transferred between circuits most efficiently if the output impedance of the first stage matches the input impedance of the following stage.

The common-emitter configuration does a good job as a general purpose amplifier. Current gain, voltage gain, and power gain are all high with this circuit configuration. The output of a common-emitter amplifier will always be phase shifted 180 degrees from the input phase. That is, when the input signal goes positive (above the dc bias level), the output signal will go negative, and vice versa. The common-emitter amplifier is probably the most commonly used of the three basic transistor amplifier configurations.

The common-base amplifier

The second basic transistor amplifier configuration is the common-base circuit, illustrated in Fig. 6-12. In this case, the input signal is fed into the transistor's emitter and the output is taken off from the collector. The base is the common element between the input and output signals. Notice that the polarity relationships between the transistor's leads remain the same as in the common-emitter configuration. The base is positive with respect to the emitter, but negative with respect to the collector. In other words, the emitter is at a negative voltage (below common ground) and the collector is at a positive voltage (above common ground). Regardless of the circuit configuration used, the transistor must be correctly biased, as explained earlier in this chapter.

Since the transistor's base lead is grounded in this circuit, its nominal value is 0 V. Resistor R_3 and capacitor C_b are placed between the base lead and the actual ground point for stability.

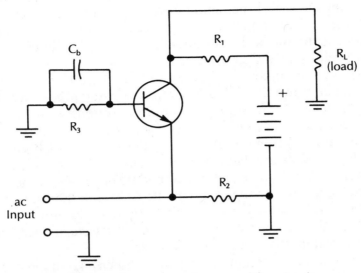

Fig. 6-12   *The common-base amplifier circuit.*

The value of both of these components are quite small to minimize the voltage drop across them. For all practical purposes, the voltage applied to the base lead of the transistor in this circuit is zero.

If you are having some trouble visualizing what is happening here, the voltage drop across resistor R_2 causes the voltage on the emitter to be below ground potential (negative). The base, as stated above, is effectively at ground potential (0 V).

The input signal varies the instantaneous voltage on the emitter lead, while the voltage on the base lead remains constant. All the transistor is concerned with is the voltage between the base and the emitter. It doesn't care which of these is varied by the input signal.

The power gain (current gain times voltage gain) of a common-base amplifier is slightly lower than a comparable common-emitter circuit using the same transistor. However, the voltage gain of the common-base amplifier is much lower than that of the common-emitter amplifier. The current gain of a common-base amplifier is negative. That is, the output current is always less than the input current.

The input and output impedances of a common-base amplifier are similar to those of a common-emitter amplifier. That is, the input impedance will be low and the output impedance will be high. The difference between the input and output impedances is significantly larger in a common-base amplifier than in a common-emitter circuit. A typical common-base amplifier circuit will usually have an input impedance of less than 100 Ω and an output impedance as high as several hundred kilohms (thousands of ohms).

While a common-emitter amplifier phase shifts the signal 180 degrees, the common-base circuit does not phase shift the signal. In other words, the output signal is in phase with the input signal. When the input signal goes more positive, so does the output signal. When the input signal goes negative, the output signal goes negative too.

The common-collector amplifier

The third basic transistor amplifier configuration is the common-collector circuit, illustrated in Fig. 6-13. Notice that this circuit employs a positive ground point. That is, all of the operating voltages in a common-collector circuit are negative. This is to maintain the correct biasing of the transistor. The collector must be at the most positive potential of the three leads. The emitter is at the most negative voltage. Resistor R_1 drops some of this negative voltage so that the base is less negative (more positive) than the emitter, but still more negative than the collector.

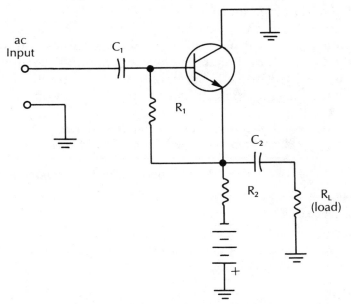

Fig. 6-13   *The common-collector amplifier circuit.*

One of the most unique features of the common-collector amplifier is that the voltage gain is always negative. In other words, the output voltage from this type of circuit is always less than the input voltage. Another way of saying this is that the voltage gain is less than unity (1).

The common-collector amplifier has fairly high current gain. The power gain (voltage gain times current gain) is somewhat positive, but relatively small, compared to the power gains of common-emitter and common-base amplifiers.

For obvious reasons, the common-collector circuit makes a pretty mediocre amplifier overall. This configuration is almost never used for actual amplification applications. It does come in handy for impedance matching applications, however. With the other two configurations, the input impedance is always considerably lower than the output impedance. In the common-collector circuit, this situation is reversed. The input impedance of a common-collector amplifier is medium to high and the output impedance is low. Remember, maximum efficiency in power transfer between circuits occurs when the impedances are matched. The output signal of a common-collector is in phase with its input. Like the common-base circuit, this configuration does not exhibit any phase shift.

The three basic transistor amplifier configurations are summarized and compared in Table 6-1. Generally speaking, a common-emitter circuit will be used in most amplification applications, except where this configuration's phase shift would be a problem. In such a case, a common-base circuit will probably be used instead. Common-collector circuits aren't normally used for actual amplification, but they are often found in

Table 6 – 1 Comparison of the three basic transistor amplifier configurations.

	Common-base	Common-emitter	Common-collector
Input impedance	Very low	Low	Medium – high
Output impedance	Very high	High	Low
Current gain	Negative	High	High
Voltage gain	Medium	High	Negative
Power gain	High	High	Low
Phase shift	0 degrees	180 degrees	0 degrees

multistage circuits for impedance matching between adjacent stages.

Darlington transistors

Bipolar transistors can sometimes become quite unstable if a high output current is required of them. A more stable high current gain can be achieved by connecting two bipolar transistors in series, as illustrated in Fig. 6-14. The emitter of transistor Q_1 is connected to the base of transistor Q_2. Both collectors are tied together. When transistors are connected in this manner, they are called a Darlington pair.

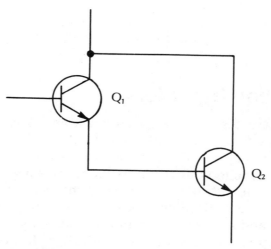

Fig. 6-14   *A more stable high current gain can be achieved by connecting two bipolar transistors in series.*

A Darlington pair can be used in most circuits almost as if they were a single "super" transistor. The current at the emitter of Q_2 is virtually the same as the current at the collectors. This allows for excellent balance between the transistors. For the best performance, the two transistors in a Darlington pair must be very closely matched.

Often, a dedicated Darlington transistor is used. This is essentially two identical transistors within a single housing. Externally such a device looks just like a regular transistor. To indicate that it is a single unit rather than two discrete (separate) transistors, a ring is drawn around the symbol, as shown in Fig. 6-15. This is one case of a semiconductor symbol where the surrounding circle is not considered optional.

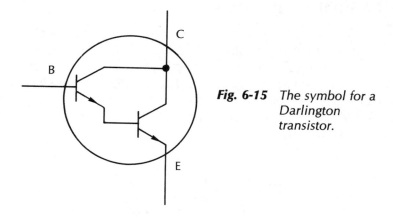

Fig. 6-15 *The symbol for a Darlington transistor.*

Unijunction transistors

So far we have discussed only bipolar transistors. There are a number of other types of transistors too. A bipolar transistor has two diodelike (PN) junctions. Another important type of transistor is the unijunction transistor. As the name suggests, this transistor has just one PN junction. This might make the component sound rather like a glorified diode. In a way, it is, except it has three leads, and a diode, by definition, has only two. The term unijunction transistor is a bit of a mouthful, so it is often abbreviated as UJT. Unijunction transistors are most commonly used in timing and oscillator circuits.

The basic internal structure of a UJT is illustrated in Fig. 6-16. Notice that this type of transistor has no collector. Instead, it has an emitter and two bases.

Most commercially available UJTs are N-type, as shown in Fig. 6-16. The two base leads are positioned at opposite ends of a relatively large chunk of N-type semiconductor. A small piece of P-type semiconductor is embedded in the middle of the main N-type section. The emitter lead is connected to this P-type semiconductor section.

The standard symbol for a UJT is shown in Fig. 6-17. Once again, the circle is often omitted, since it adds no extra informa-

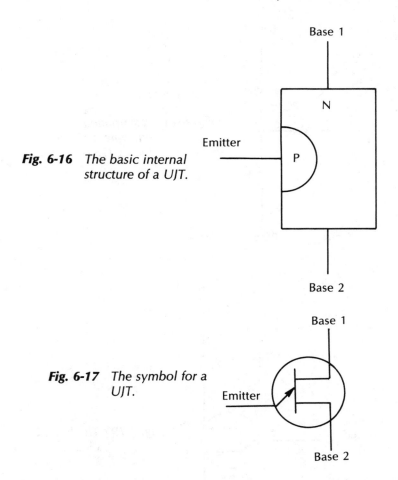

Fig. 6-16   *The basic internal structure of a UJT.*

Fig. 6-17   *The symbol for a UJT.*

tion to the symbol. It just makes the schematic diagram look a little neater and easier to read.

Electrically, the N-type section of the UJT acts like a simple two-stage resistive voltage divider with a diode (the single *PN* junction) connected to the common ends of the two resistances. Figure 6-18 shows a very simplified equivalent circuit for a UJT.

The basic UJT circuit is illustrated in Fig. 6-19. A voltage is applied between base 1 and base 2. This voltage reverse biases the diode. Of course, this means that no current can flow from the emitter to either base. The UJT is cut off.

Now, let's suppose there is an additional variable voltage source connected between the emitter and base 1. This is the UJT's input signal. As this emitter-base 1 voltage is increased from zero, a point will be reached when the internal "diode"

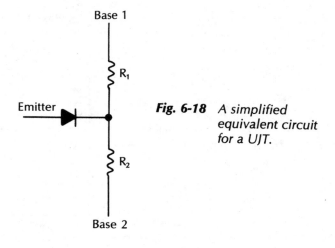

Base 1

R_1

Emitter

R_2

Base 2

Fig. 6-18 *A simplified equivalent circuit for a UJT.*

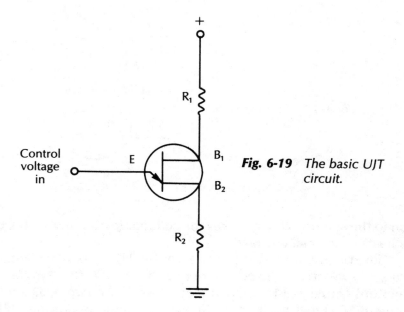

+

R_1

Control voltage in

E B_1

B_2

R_2

Fig. 6-19 *The basic UJT circuit.*

(the *PN* junction) will become forward biased. Beyond this point current can flow between the emitter and the bases.

In many practical circuits using UJTs, the exact type often isn't particularly critical. Except in critical applications, there is usually ample room for substitution of different UJTs. In some applications, substituting a different UJT for the specified type may affect some operating parameters in the circuit. Timing periods and oscillator frequencies could be affected in some cases. Of course, if you make a substitution for any UJT, you must make sure that the replacement transistor is also a UJT. A bipolar transistor or an FET (discussed in the next section) certainly won't work in a UJT circuit.

P-type UJTs are available, though they are far less common than N-type UJTs. In a P-type UJT, the N-type and P-type semiconductor sections are reversed. All circuit polarities are also reversed. This reversal is reflected in the direction of the emitter arrow in the symbol for a P-type UJT, as illustrated in Fig. 6-20. Naturally, if you make a substitution for a UJT, you can't replace a P-type device with an N-type unit, or vice-versa.

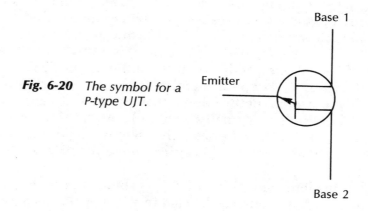

Fig. 6-20 *The symbol for a P-type UJT.*

FETs

Another increasingly important type of transistor is the field-effect transistor (FET). Transistors were originally developed as solid-state equivalents to triode vacuum tubes. A solid-state component is smaller, less fragile, and more reliable than the glass tube. It also operates with much greater efficiency and less heat. However, the bipolar transistor's operation doesn't quite correspond to that of a triode vacuum tube. For one thing, the bipolar transistor is a current amplifier, while the vacuum tube is a voltage amplifier. In many applications, this ultimately doesn't make much difference, although it will be reflected in differences in the circuit design. However, in some applications, the operating characteristics are highly desirable or even essential, but the bulk, excess heat, and power consumption of vacuum tubes are not so desirable. Electronic component technicians eventually developed a solid-state component that did a pretty good job of mimicking the operation of a triode vacuum tube. This particular solid-state component is known as the field-effect transistor (FET).

An FET, like a bipolar transistor, has three leads, but these leads serve somewhat different functions, so they are assigned different names. The three leads of an FET are called the source, the gate, and the drain.

Both N-type and P-type FETs are available, although P-type FETs are relatively rare. When no semiconductor type is specified, it is usually safe to assume that the FET in question is an N-type device. For the time being we will focus our attention solely on the more common N-type FET. Of course, P-type FETs work in exactly the same way, except all polarities are reversed.

The basic structure of an FET is illustrated in Fig. 6-21. Like the UJT, the main body of this component is a single, continuous length of N-type semiconductor material. But in the FET, there is a small section of P-type material placed on either side of the main N-type section. Both of these small P-type sections are electrically tied together internally. The lead connected to these two P-type sections is called the gate. The other two leads (the source and the drain) are connected to either end of the piece of N-type material.

The standard symbol for an FET is shown in Fig. 6-22. The leads of an FET will often be labeled in a schematic, because it is difficult to distinguish the source from the drain just by looking at the symbol. Still, the lead labels are omitted in many schematic diagrams.

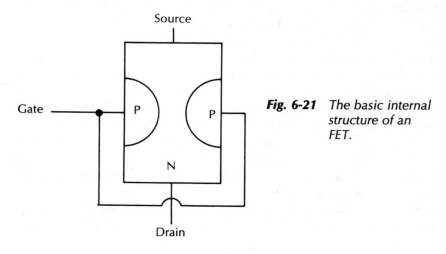

Fig. 6-21 *The basic internal structure of an FET.*

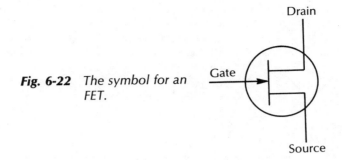

Fig. 6-22 *The symbol for an FET.*

The best way to describe the operation of an FET is to use a mechanical analogy. Imagine that we are pouring water through a tube or pipe. The water is poured into the system at the source and flows out of the drain. A movable valve in the pipe is the gate. When the valve (gate) is opened, as illustrated in Fig. 6-23, water can flow freely through the pipe (from source to drain). If, on the other hand, the valve is partially closed, as in Fig. 6-24,

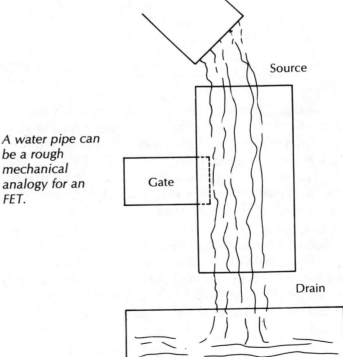

Fig. 6-23 *A water pipe can be a rough mechanical analogy for an FET.*

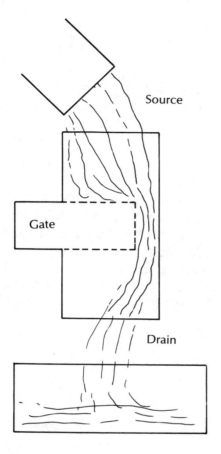

Fig. 6-24 *Less water flows through the pipe when the valve is partially closed.*

the amount of water that can flow through the pipe is limited; less water comes out of the drain. Quite similarly, the gate terminal of an FET controls the amount of electric current that can flow from the source lead to the drain lead.

For proper operation, an FET must be biased as shown in Fig. 6-25. A negative voltage applied to the gate lead reverse biases the *PN* junction producing an electrically charged region within the N-type material. This electrically charged region is called an electrostatic field.

This electrostatic field opposes the flow of electrons through the N-type section, acting somewhat like the partially closed valve in our mechanical model. The higher the negative voltage applied to the gate, the larger the electrostatic field, and the lesser the current that is allowed to pass through the FET from source to drain.

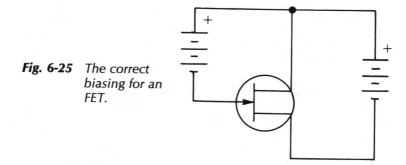

Fig. 6-25   *The correct biasing for an FET.*

The current path from the source to the drain is sometimes called the channel. The voltage applied to the gate, effectively controls the "size" (current capacity) of the channel.

The operation of an FET is quite analogous to the functioning of a triode vacuum tube. The FET's gate corresponds to the tube's grid, controlling the amount of current flow. The FET's source is the equivalent of the tube's cathode, acting as the source of the electron stream through the device. The FET's drain serves the same basic function as the tube's plate, draining off the electrons from the component.

FETs have a very high input impedance, so they draw very little current. This means they place a minimal load on their input source. This connection between the input impedance and the current draw is due to Ohm's law. Since $I = E/Z$, increasing Z (impedance) without changing E (voltage) will decrease I (current).

Because of their very high input impedance, FETs can be used in highly sensitive measurement applications and in any other circuit where it is important to avoid overloading (drawing heavy currents from) previous circuit stages. Bipolar transistors have much lower input impedances than FETs.

A P-type FET is similar to the N-type FET we have been discussing in this section except all polarities are reversed. To indicate this change in polarity, the symbol for a P-type FET, shown in Fig. 6-26, has the arrow reversed in the gate.

Actually, while we have been using the generalized name FET in this section, we have only been looking at one particular kind of FET. There are others. The sort of FET we have been dealing with is more specifically called a junction FET. This is commonly abbreviated as JFET or J-FET.

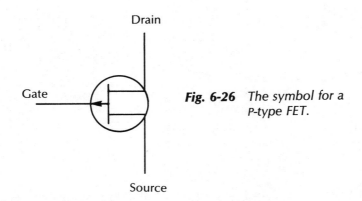

Fig. 6-26 *The symbol for a P-type FET.*

MOSFETs

Another type of FET does not have an actual PN junction. The name for such a device is the insulated gate field-effect transistor (IGFET) because the gate is insulated from the channel (source-to-drain current path). Most commonly, this insulation is achieved by using a thin slice of metal as the gate instead of the usual piece of semiconductor crystal. This slice of metal is oxidized on the side that is placed against the semiconductor channel. Metal oxide is a very poor conductor, so this layer of oxidization effectively insulates the gate. When a metal oxide is used to insulate the gate, the IGFET is often called a metal-oxide silicon field-effect transistor (MOSFET).

The semiconductor channel is backed by a substrate of the opposite type of semiconductor. That is, if the channel is made of an N-type semiconductor, a P-type semiconductor is used as the substrate. IGFETs with P-type semiconductor channels are considerably less common.

The basic structure of an N-type MOSFET is shown in Fig. 6-27. Figure 6-28 shows the symbol for an N-type MOSFET. Sometimes the surrounding circle may be omitted. This doesn't change the meaning of the symbol. Notice that this kind of FET has a fourth lead connected to the substrate. This fourth lead isn't used in all circuits. Often, the substrate lead is simply shorted directly to the source lead.

To indicate a P-type IGFET, all that has to be done is to reverse the direction of the arrow, which in this case is shown in the substrate. The symbol for a P-type IGFET appears as Fig. 6-29.

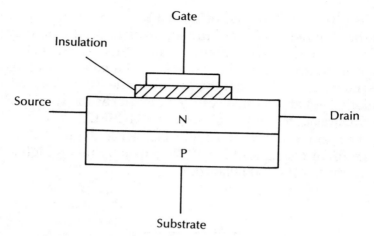

Fig. 6-27 *The basic structure of an N-type MOSFET.*

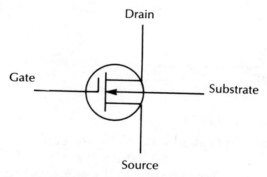

Fig. 6-28 *The symbol for an N-type MOSFET.*

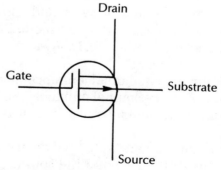

Fig. 6-29 *The symbol for a P-type IGFET.*

Even though the gate of an IGFET is physically insulated from the channel, it can still induce an electrostatic field into the nearby semiconductor channel, and this electrostatic field (which is controlled by the voltage on the gate) can be used to limit the current flow from the source to the drain. The substrate is usually kept at the same voltage potential as the source. Figure 6-30 illustrates the proper biasing of a MOSFET.

The drain resistance of a typical IGFET is about 50 kΩ (50,000 Ω). The drain resistance for a JFET is usually much higher, typically about 1 MΩ (1,000,000 Ω).

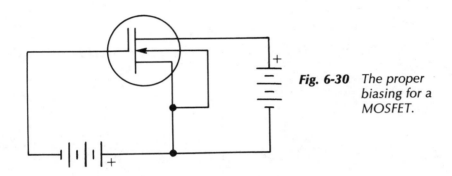

Fig. 6-30 *The proper biasing for a MOSFET.*

Enhancement-mode FETs

Both JFETs and insulated gate FETs are depletion-mode FETs. This means that these devices reduce the current flow by increasing the negative voltage applied to the gate (assuming an N-type channel).

Another kind of FET operates in the enhancement mode. In an enhancement-mode FET, the source and the drain are not parts of a continuous channel of semiconductor material. The basic structure of a typical enhancement-mode FET is illustrated in Fig. 6-31. Enhancement mode FETs always have an insulated gate.

The symbol for an N-channel enhancement-mode FET is shown in Fig. 6-32. For a P-channel enhancement-mode FET, the direction of the arrow in the substrate line of the symbol is reversed, as shown in Fig. 6-33.

In operation (assuming an N-channel device), a positive voltage is applied between the gate and the source. The higher this voltage is, the greater the number of holes drawn from the N-type source into the P-type substrate. These holes are then drawn into

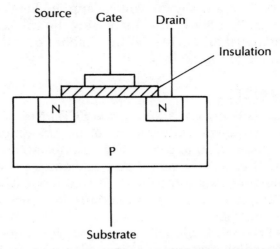

Fig. 6-31   *The basic structure for a typical enhancement-mode FET.*

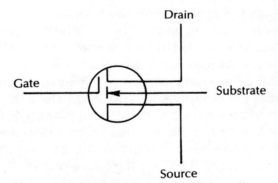

Fig. 6-32   *The symbol for an N-channel enhancement-mode FET.*

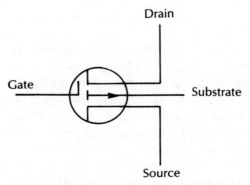

Fig. 6-33   *The symbol for a P-channel enhancement-mode FET.*

the N-type drain region by the voltage applied between the drain and the source. In other words, increasing the voltage on the gate increases (rather than decreases) the amount of current flowing from the source to the drain.

Other types of FETs

A lot of work is being done to improve FETs. New kinds of FETs are announced by semiconductor manufacturers almost every year. It would be impractical to try to fully describe all available kinds of FETs here. Most of the newer FET categories are variations on the basic depletion-mode and enhancement-mode IGFETs already discussed. The new types can essentially be considered higher-grade replacements.

Some of the kinds of FETs you might encounter include the V-MOSFET, which features a V-shaped groove to increase the power-handling capability, and the COMFET (conductivity-modulator field-effect transistor).

The V-MOSFET is capable of very high switching speeds, even into the nanosecond range. Besides its higher power-handling capability, this device has the advantage of an extremely low "on" resistance—as low as a few tenths of an ohm.

The COMFET also offers an extremely low drain resistance — typically just 0.1 Ω or less. A COMFET can also withstand much higher power levels than a comparable standard IGFET. A typical COMFET can block up to 400 V when forward biased and as much as 100 V when reverse biased. Functionally, a COMFET behaves as if it was a heavy-duty IGFET feeding into a pair of direct-coupled bipolar transistors. This is illustrated in the equivalent circuit shown in Fig. 6-34.

SCRs

Another special purpose semiconductor device is the silicon controlled rectifier (SCR). I suppose that it is somewhat debatable whether an SCR is a true transistor. SCRs are not normally used for amplification, but for switching applications. However, transistors are often used in switching circuits too.

The SCR belongs to a class of components known as thyristors, a contraction of thyratron transistor. The shorter term (thyristor) is the one most commonly used in modern technical literature. Sometimes thyristors are called reverse-blocking thy-

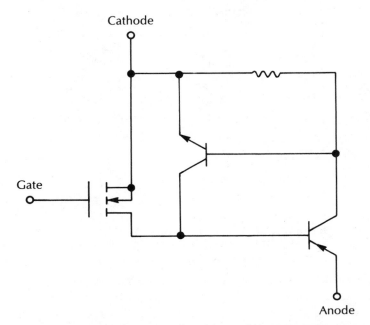

Fig. 6-34 *The equivalent circuit for a COMFET.*

ristors, but this terminology is redundant because all thyristors are, by definition, reverse blocking.

Generally, thyristors are used to control the power fed to a load. In the early days of electronics, such applications were performed by large resistors and rheostats, and bulky, expensive thyratron tubes. In actual practice, thyratron tubes were seldom used unless they were absolutely necessary. It was more convenient, economical, and practical to avoid them if at all possible.

Semiconductor thyristors, on the other hand, are quite compact and relatively inexpensive. They are also highly efficient and can often handle very high currents. Some thyristor devices are rated for current-handling capacities as high as 4000 A. Because thyristors are so small and reliable, they are often a very attractive alternative to relays or other electromechanical switching devices.

Thyristors are made up of four semiconductor sections of alternating type, as shown in Fig. 6-35. Notice the strong similarity to the four-layer diode discussed in chapter 5. The difference between a thyristor and a four-layer diode is that a thyristor usually has three leads. Some technicians might consider the four-layer diode a sort of thyristor.

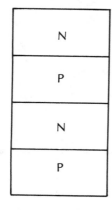

Fig. 6-35 *Thyristors are made up of four semiconductor sections.*

There are several different types of thyristors. The differences between the thyristor types lie in the placement of the leads. The SCR is probably the most commonly used type of thyristor, and it is one of the easiest to understand.

In many ways, the SCR is functionally closer to the diode than to the transistor. This is reflected in the standard symbol for the device, shown in Fig. 6-36. As with most of the components in this chapter, the surrounding circle may be omitted without altering the meaning of the symbol. Notice that the symbol for the SCR is basically the same as the symbol for a semiconductor diode, with the addition of a third lead.

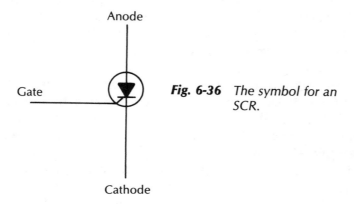

Fig. 6-36 *The symbol for an SCR.*

The two main leads to this component are called the anode and the cathode, just as with an ordinary diode. The third lead is called the gate. The leads of an SCR usually are not labeled in schematic diagrams because they are indicated by the symbol itself.

The operation of this device is implied by its name—silicon controlled rectifier. The main body of the component is made of silicon. Rectifier means the device is used for rectification, or diode action. The important term is the one in the middle—controlled. The rectification process of an SCR can be electrically controlled by an external signal.

The structure of an SCR is illustrated in Fig. 6-37. Notice that this device is quite similar to the four-layer diode (discussed in chapter 5), except that a third lead (the gate) has been added.

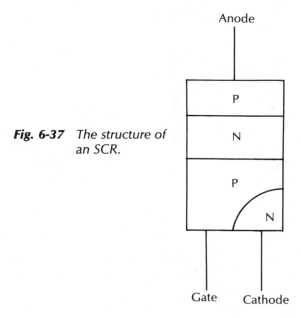

Fig. 6-37   *The structure of an SCR.*

To get an idea of how this component works, first consider the simple (nonfunctional) circuit shown in Fig. 6-38. Here we have a forward-biasing voltage applied between the cathode and the anode. Because the gate is grounded, its potential is 0 V. If this was a regular diode, current would flow. But under these conditions, the SCR continues to block the flow of current, just as if it was reverse biased. This will always be the case, as long as the gate is grounded.

Now, let's apply a variable voltage to the gate terminal of the SCR, as illustrated in Fig. 6-39. If we start the gate voltage at zero and gradually increase it, at some point it will exceed a specific trigger voltage. The actual trigger voltage will depend on the design of the individual SCR used. When this trigger voltage is

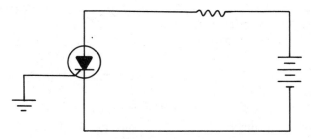

Fig. 6-38   *This nonfunctional circuit is used to illustrate the operation of an SCR.*

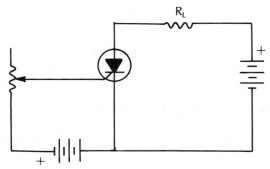

Fig. 6-39   *The circuit of Fig. 6-38 with a variable voltage applied to the gate of the SCR.*

exceeded, current will start to flow through the SCR from the cathode to the anode against only a small internal resistance, as with an ordinary forward-biased diode.

Once started, this current will continue to flow—even if the voltage on the gate is removed. The gate can turn the SCR on, but it can't turn it back off. The only way to stop the current flow once it has started is to decrease the positive voltage on the anode to a very low level or remove it altogether. Reverse biasing the SCR will certainly do the trick. When the anode voltage drops below a predetermined level (which depends on the internal characteristics of the SCR, and sometimes certain external circuitry), the current flow will be blocked. It will stay blocked, even if the anode voltage returns to its original full value. The SCR will not conduct current until it is turned back on by an appropriate voltage pulse on the gate lead.

A special type of diode, called a trigger diode, is often used in conjunction with an SCR. The trigger diode is connected in series with the gate, as shown in Fig. 6-40. This trigger diode is

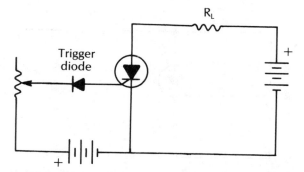

Fig. 6-40 *A trigger diode is placed in series with the gate of an SCR.*

ordinarily reverse biased, and it is designed to exhibit a sharp pulse when its trigger voltage is exceeded. This is shown graphically in Fig. 6-41. Because of the sharp, clean triggering pulse it provides, use of a trigger diode can result in cleaner and more precise and reliable triggering of the SCR.

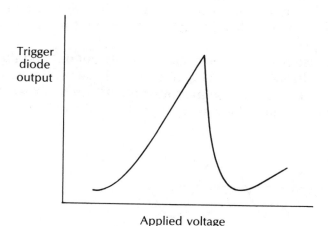

Fig. 6-41 *The trigger diode puts out a sharp pulse when its trigger voltage is exceeded.*

The most common application of an SCR is to control ac voltages. SCR control circuits are used in light dimmers and motor speed controllers, among other applications. A very simple circuit of this type is illustrated in Fig. 6-42. When the voltage applied to the trigger diode reaches a specific point in the ac cycle, it turns the SCR on with a sharp voltage pulse. Later in the cycle, the input voltage starts to decrease. When the applied voltage to the anode drops below the holding level of the SCR, the

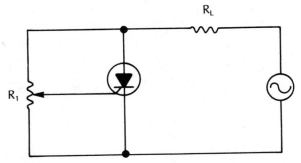

Fig. 6-42   *A very simple ac controller circuit using an SCR.*

current flow is cut off until the point in the next cycle when the trigger diode emits a new voltage pulse.

Figure 6-43 illustrates the input signal and several possible output signals for this type of circuit. Varying the resistance value of R_1 will alter the voltage applied to the trigger diode and, thus, the point in the cycle when the SCR is turned on. By controlling the switch-on point in this cycle, we can control how much of each complete input cycle will get through to the output.

Because part of the ac cycle is simply cut out and the instantaneous voltage remains at zero during those off times, the average value of the output signal must obviously be lower than that of the input cycle. The less time per cycle the current is allowed

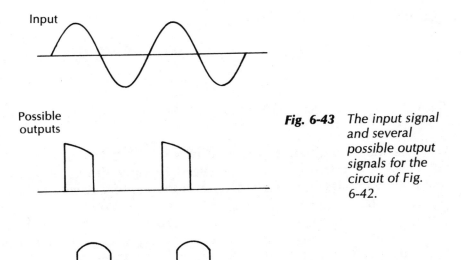

Input

Possible
outputs

Fig. 6-43   *The input signal and several possible output signals for the circuit of Fig. 6-42.*

to flow through the SCR, the lower the effective value of the output voltage.

Diacs and triacs

SCRs and trigger diodes are, by definition, single-polarity devices. They can conduct current only in a single direction. They can only conduct, at most, for one-half of each ac cycle. Of course, this means that all of the input energy in the other half-cycle is wasted and dissipated as heat. In some applications, this isn't much of a problem. But in other applications, it would be highly desirable to have a bidirectional (nonpolarized) SCR or trigger diode.

A diac is a dual-trigger diode that is bidirectional. It will produce an output pulse on each half-cycle. In effect, a diac functions like a pair of back-to-back trigger diodes, connected as shown in Fig. 6-44. The upper "diode" conducts during the positive half-cycles, while the lower "diode" conducts during the negative half-cycles, so the entire input cycle is covered.

Fig. 6-44   *A diac functions like a pair of back-to-back trigger diodes.*

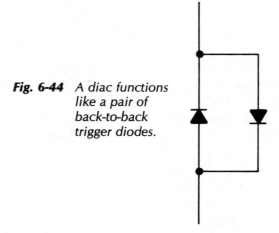

The symbol for a diac is shown in Fig. 6-45. Notice how this symbol suggests two back-to-back diodes of opposite polarity.

Similarly, a triac is a dual, bidirectional SCR. It functions like two separate SCRs that are connected in parallel, but in opposite directions, as illustrated in Fig. 6-46. Notice that only one of these two SCRs can conduct at any given instant. When one is conducting, the other is cut off. This is why their gates are

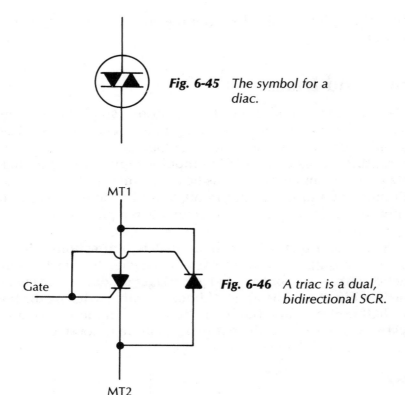

Fig. 6-45 *The symbol for a diac.*

Fig. 6-46 *A triac is a dual, bidirectional SCR.*

tied together. Only a single gate signal is required to operate both SCRs. If a gate pulse is received while SCR A is forward biased, it will start to conduct. If SCR A is forward biased, then SCR B must be reverse biased, so SCR B will ignore the trigger pulse on its gate and remain off. Of course, at any particular instant both SCRs might be cut off, but there will never be more than one on at any given instant. A simplified drawing of the construction of a triac is shown in Fig. 6-47.

The symbol for a triac is shown in Fig. 6-48. Notice that this is basically the same as the diac symbol with the addition of the gate lead. Notice how this symbol suggests two back-to-back diodes of opposite polarity. Figure 6-49 shows a simplified equivalent circuit for a triac.

The control terminal on a triac is called the gate, as with an SCR. But the other two terminals require new names. Because the triac is nonpolarized, it wouldn't make sense to call the two end leads the anode and the cathode. Some technicians call the two main leads of a triac anode 1 and anode 2, or A1 and A2.

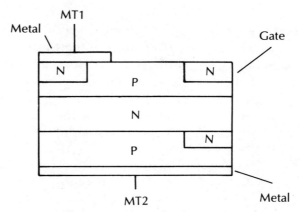

Fig. 6-47   *The construction of a triac.*

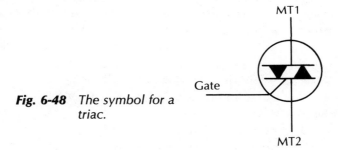

Fig. 6-48   *The symbol for a triac.*

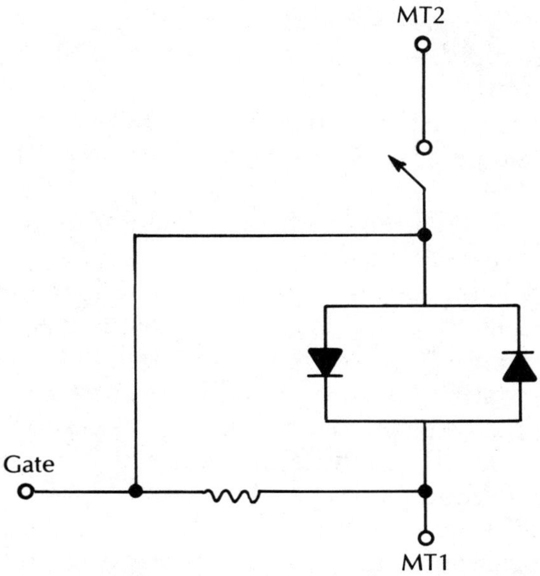

Fig. 6-49   *A simplified equivalent circuit for a triac.*

More commonly, these leads are labeled MT1 and MT2 for main terminal 1 and main terminal 2. Personally, I think this terminology makes a lot more sense than identifying these nonpolarized terminals as anodes, which implies a positive polarity.

Because of its bidirectional current-carrying capabilities, the triac is a very versatile device. There are four possible combinations for triggering a triac. We call these the triggering modes. The differences in the four triggering modes lie in the relative polarities of the leads.

Usually, all current and voltage polarities for a triac are given with respect to MT1. This convention is followed simply as a matter of convenience and to avoid unnecessary confusion. There is nothing particularly special about MT1. In fact, MT1 and MT2 are functionally identical and more or less interchangeable. MT1 is just an arbitrarily chosen common reference point. The triac's four standard triggering modes are as follows:

- Mode A
  —MT2 positive with respect to MT1
  —Gate pulse positive with respect to MT1
- Mode B
  —MT2 positive with respect to MT1
  —Gate pulse negative with respect to MT1
- Mode C
  —MT2 negative with respect to MT1
  —Gate pulse positive with respect to MT1
- Mode D
  —MT2 negative with respect to MT1
  —Gate pulse negative with respect to MT1

Each of these triggering modes has a different current requirement to trigger the triac.

Mode A is the easiest to trigger, and it is therefore the most commonly used mode. This triggering mode has the lowest current requirement. The gate current required to trigger the triac in Mode A is identified as I_{gt}. This value is determined by the internal characteristics of the particular triac being used. The manufacturer will usually include the nominal I_{gt} value in the spec sheet for the component.

In Mode B the triac isn't nearly as efficient as it is in mode A. A gate current of at least five times I_{gt} is required to trigger the triac in mode B.

Mode C and mode D are basically similar to one another. A gate current of about twice I_{gt} is needed to trigger the triac in each of these modes, regardless of whether the gate signal is positive or negative with respect to the MT1 terminal.

A simple ac control circuit using a diac and a triac is illustrated in Fig. 6-50. Notice that this is essentially the same circuit we used with the regular, single-polarity trigger diode and SCR

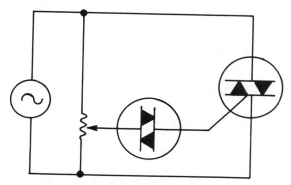

Fig. 6-50 *A simple ac control circuit using a triac and a diac.*

in the preceding section. The difference here, of course, is that this circuit is operative in both half-cycles of the input ac waveform. This means less of the input power is wasted; more of the input power can reach the circuit's output. In other words, this triac circuit is much more efficient than the SCR version. Some typical input and output signals for this circuit are shown in Fig. 6-51.

For convenience and clarity, we will discuss this circuit as if two back-to-back trigger diodes and SCRs were used. Of course, both trigger diodes are contained within the diac, and both SCRs are contained within the triac. When point A is reached in the cycle, trigger diode A (within the diac) conducts, triggering SCR A (within the triac). At point B, SCR B is turned off. Trigger diode B and SCR B are both inactive throughout this entire half-cycle.

During the second half-cycle, trigger diode A and SCR A are inactive, while trigger diode B and SCR B come into play. At point C in the input cycle, SCR B is turned on by a trigger pulse from trigger diode B. Then, at point D, SCR B is cut off once again. Then the entire cycle is repeated. Trigger diode A and SCR A alternate with trigger diode B and SCR B. The output signals of

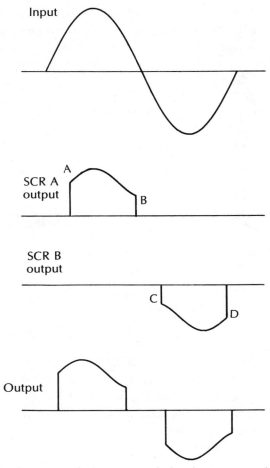

Fig. 6-51 *Typical input and output signals for the ac control circuit of Fig. 6-50.*

the two SCRs within the triac are combined into a single waveform that will be applied as the power source to the load.

Notice that the output of the triac always removes some of the input waveform, so the power at the circuit's output will always be less than at the input. The only question is how much of the input power will be diverted by the time it reaches the output. This is controlled by setting the triggering points (points A and C) by adjusting the voltage to the diac (trigger diodes) with the potentiometer. The output power can be smoothly varied over a range extending from near zero to a little less than the original input power.

Quadracs

Because diacs and triacs are almost always used together, they are sometimes combined into a single component called a quadrac. The symbol for a quadrac is shown in Fig. 6-52. The surrounding circle is almost always used in this case to indicate that this is a single component, not a separate diac and triac.

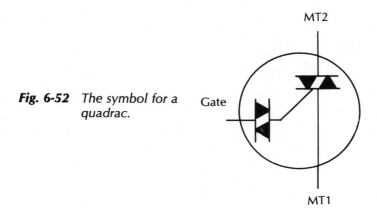

Fig. 6-52   *The symbol for a quadrac.*

The leads on a quadrac are labeled in the same way as on a triac. That is, the three terminals are called MT1, MT2, and gate. The internal diac is at the gate terminal, of course.

Currently, quadracs are quite expensive and rather rare, especially on the hobbyist market. You might encounter them occasionally if you work on industrial equipment.

❖ 7
Linear ICs

AS ELECTRONICS TECHNOLOGY INCREASES, CIRCUIT SIZE HAS decreased. More power can be packed into a smaller space. Many modern devices would be impractically expensive and physically unwieldy, if not outright impossible, if not for the many recent advances in miniaturization. The switch from tubes to transistors was the first major step forward in this direction.

Once circuit designers saw the advantages of the more compact solid-state technology, they wanted to compress their circuits even more. Much of the push for miniaturization came from the space program. On a satellite, everything must be as small and as lightweight as possible.

Several manufacturers experimented with modules. These were small generalized circuits, hardwired together in as small a space as possible. The entire module was enclosed in epoxy, with only the power supply, input, and output leads extending out of the body of the unit. The module approach worked, but it was severely limited in its applications, and the savings in space weren't always very great. Moreover, there was often a serious problem with heat dissipation from the closely packed components.

Electronics miniaturization really came into its own with the invention of the integrated circuit (IC). An entire circuit can be etched into a small slab of semiconductor crystal and enclosed in a single, compact housing.

An IC is made up of a number of tiny silicon wafers that are specially treated to simulate a variety of separate transistors, diodes, resistors, and capacitors. A tiny package that is smaller than a dime can take the place of a couple of dozen (or more) discrete components. Integrated circuits are often called chips or IC chips because the internal circuitry is etched directly onto a chip of semiconductor crystal. This is usually silicon, but other semiconductor materials are used occasionally.

Levels of integration

There are several possible levels of integration, depending on the complexity of the circuit simulated by the IC. The standard levels of integration are (from lowest to highest):

- SSI—small-scale integration,
- MSI—medium-scale integration,
- LSI—large-scale integration, and
- VLSI—very large-scale integration.

SSI devices are simple arrays of closely matched components or relatively simple circuits that can be built around a handful of ordinary discrete components. The old epoxy modules discussed earlier were equivalent to SSI ICs. While not as impressive as the higher levels of integration, an SSI IC represents a significant reduction in circuit size and usually a reduction in cost too. SSI ICs are used as basic building blocks in more complex systems. Because of their inherent simplicity, SSI devices tend to be very versatile and can be employed in a great many different applications. SSI ICs are the most commonly used type of IC.

LSI devices, on the other hand, include much more complicated circuitry than SSI ICs. Sometimes a single LSI IC can take the place of literally hundreds or even thousands of discrete components. Naturally, this represents a very substantial reduction in the circuit's size and weight, and usually in the cost as well.

LSI ICs are designed for very specific and specialized functions. They usually can't be used in many different applications. If you open up a pocket calculator, you'll find that most of the circuitry is contained within a single IC. This IC couldn't be used for much else, but it makes an efficient and compact calculator. The IC used in a pocket calculator is an example of an LSI device.

Between these two extremes of integration (SSI and LSI) are MSI devices. An MSI IC is more complex than a SSI unit, but not as complex as an LSI device. MSI ICs are usually designed to perform some specific function within a larger system. For example, an audio amplifier circuit might be replaced by an MSI chip, usually along with a handful of external components.

Recently IC manufacturers have improved their miniaturization techniques enough to develop VLSI ICs. A VLSI device is even more complex and specialized than an LSI unit.

The first commercial IC, which appeared back in 1961, included just four on-chip bipolar transistors. A modern LSI device, such as Motorola's 68000 CPU (central processing unit—used in computers) IC might contain 65,000 to 70,000 on-chip transistors. Some of the newer VLSI ICs now being designed have as many as 250,000 on-chip transistors.

IC packaging

Integrated circuits come in a number of standardized package types. Some are enclosed in round, plastic or metal cans, looking somewhat like rather oversized transistors, except there are usually more than three leads. Round can ICs usually have 8, 10, or 12 leads.

Most modern ICs, however, are housed in dual-inline packages (DIPs). The ICs shown in Fig. 7-1 are all standard DIP housings. These rectangular plastic packages, usually black in color, have two parallel rows (or lines) of leads or pins, hence the name. According to some sources, DIP stands for dual-inline pins.

DIP ICs are most commonly found with 8, 14, or 16 pins. A few LSI and VLSI devices might have even more pins (24-, 28-, and 40-pin DIPs are not uncommon). But the vast majority of ICs you are likely to encounter will have 8, 14, or 16 pins. Figure 7-1 shows how these pins are numbered. The same basic approach is also used to number the pins on larger DIPs.

There is a notch or a circle etched into one end of the ICs casing to help identify pin 1. This mark will always be at the front of the device, with pin 1 to its left, when looking at the IC from above and holding it so the mark is at the top.

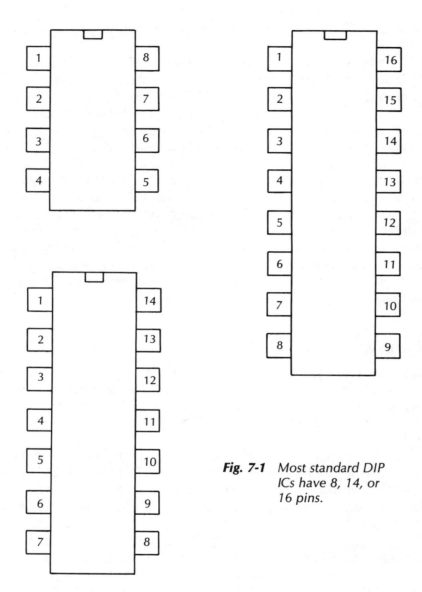

Fig. 7-1 *Most standard DIP ICs have 8, 14, or 16 pins.*

Representing ICs in schematics

For simple arrays, usually the standard symbols for the miniaturized components will be used, with a dotted box surrounding these symbols to indicate they are part of a single IC device. A simple four-transistor array is illustrated in Fig. 7-2. Notice that the optional rings around the transistor symbols are never used for on-chip transistors in an IC.

NC

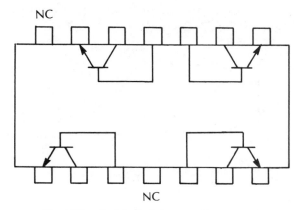

Fig. 7-2 *A simple four-transistor array.*

Most ICs are too complex to show their internal components individually. Further, there would be no point in doing so. It would just clutter and overcomplicate the schematic. In practice, ICs are treated as "black boxes." Who knows or cares what's actually in the black box. As long as you put in the right input signal(s), you'll get the correct output signal(s) from it. Therefore, most ICs are shown in schematic diagrams as boxes. The leads are numbered for identification, but they don't necessarily have to be drawn in numerical order. In the schematic, the leads can be arranged into any convenient pattern for the greatest clarity. A typical example of the way an IC looks in a schematic diagram is shown in Fig. 7-3.

A few types of ICs have special symbols that indicate their function. This is most often the case with SSI digital ICs, which will be discussed in chapter 8. One example of a special symbol

Fig. 7-3 *ICs are usually indicated in schematic diagrams as a simple box.*

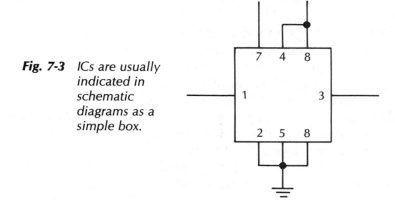

for a linear IC is an amplifier. Amplifiers are often drawn as a triangle, as shown in Fig. 7-4.

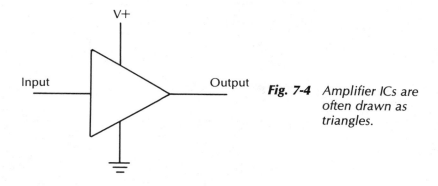

Fig. 7-4 *Amplifier ICs are often drawn as triangles.*

IC sockets

Because of the many closely spaced pins, soldering and desoldering ICs can be something of a problem. Like all semiconductors, too much heat can quickly damage or destroy an IC chip. To prevent such overheating problems, IC sockets are often used. The socket is soldered into place in the circuit, and the actual IC chip is inserted into the socket only after all of the soldering has been completed. Some electronics technicians and hobbyists swear by IC sockets and others swear at them.

One common criticism against the use of IC sockets is that it is all too easy to insert the IC into the socket backwards. When power is applied, the chip will probably be ruined. I don't feel this argument amounts to very much. It is just as easy to install the IC backwards when soldering it into the circuit. Carelessness is carelessness. Don't blame it on the socket.

It is true, however, that in some circuits, the socket may not provide a good electrical connection. This is most likely the case when very high-frequency (VHF) radio signals are involved. For at least 90% of all practical electronics work, there will be no noticeable difference in the performance of a socketed IC and an unsocketed IC. Also, in portable equipment, there might sometimes be a problem with an IC getting bounced out of its socket.

When inserting an IC into a socket, it is extremely important to make sure all of the chip's pins are straight and that they actually slip into the socket holes. Sometimes a pin might get bent up under the body of the IC rather than going into its socket hole. Of course, in this case, there is no electrical connection to that pin and the IC probably will not function as desired.

Probably the strongest argument against the use of IC sockets is that they are expensive. IC sockets typically cost about fifty cents to one dollar apiece. Many SSI and MSI ICs actually cost less than this. Some electronics technicians and hobbyists argue that it doesn't make sense to use a socket that costs more than than the chip it is intended to protect. On the other hand, mistakes and accidents happen. If the IC ever has to be replaced for any reason, the job will be a lot easier if a socket was used. You could consider the cost of the sockets as a cheap insurance policy against frustration and the time wasted desoldering and resoldering.

Linear versus digital

New IC devices are being developed almost every day—especially MSI and LSI units. It would clearly be beyond the scope of this book to discuss all available ICs. Instead we will concentrate on some of the more common types of "building block" devices and general categories of ICs. For the most part, we will be dealing with SSI and MSI chips.

There are two major and mostly incompatible categories of ICs. Some ICs are linear, or analog devices, while others are digital devices. For our purposes, linear and analog mean pretty much the same thing, and in this context, the terms are interchangeable.

In a linear circuit, signals can take any of a continuous range of values that can be graphed as a line. In audio systems, for example, the electrical signal is a direct analog of the acoustic signal. An additional signal value can always be squeezed between any two adjacent signal values in a linear system.

On the other hand, in a digital circuit, all signals must take on very discrete, noncontinuous values. The signals are thought of as numbers, or digits. Only whole numbers can be used. For example, the signal has to be either 7 or 8. There is nothing in between. A digital system does not permit 7.5 or 7.75.

In this chapter we will look at linear ICs. Digital ICs will be covered in chapter 8.

Until the 1970s, virtually all electronic circuits were analog.

Linear ICs cover a very, very wide range of applications and circuit types. A few of the more common linear functions are outlined in Table 7-1. We will look at just a few of these functions and how they are performed by linear ICs.

Table 7 – 1   Some typical linear IC functions.

Amplifiers:
☐ Audio amplifiers
☐ rf Amplifiers
☐ Operational amplifiers
Voltage regulators
Filters
rf Tuners
Modulators
Oscillators
Signal generators
Signal comparators
Timers
Electronic switches

Arrays

The simplest integrated circuits are arrays. An array is just several closely matched components etched onto a single semiconductor chip. Most arrays contain multiple transistors, but diode and resistor arrays are not uncommon.

The chief advantage of using an array rather than discrete components is that the array IC takes up less space. In some cases, an array may be less expensive than separate discrete components. Since all of the components in an array IC are formed from the same piece of silicon, they tend to be very closely matched in all operating characteristics. In some critical applications it might be difficult to achieve such close matching with separate discrete components.

In some array ICs, the leads to the various individual components may be brought out independently. All circuit connections must be made externally in the circuitry built around the IC. Other array ICs feature certain internal connections between the on-chip components. In some cases, there may be common connections. For example, all diode anodes might be hardwired to a common lead, while the cathodes are brought out individually to independent IC pins. Sometimes, the on-chip components may be hardwired into certain common circuit patterns. On-chip transistors might be connected into Darlington pairs, or they might be arranged for convenient use in common-base or common-emitter amplifier circuits.

Amplifiers

Perhaps the most basic of all electronic functions is amplification. Given this, it is not surprising that a great many amplifier ICs have appeared on the market over the years. Amplifier circuits surely outnumber any other type of analog circuit in practical electronics work. Almost every electronic system or circuit of any complexity includes at least one (usually more) amplifier stage.

In very simple terms, an amplifier is a circuit with gain. It increases the amplitude (strength) of an electrical signal. The output signal from an amplifier is a direct analog of the input signal, except that it is now larger. Actually, not all amplifier circuits really boost the signal level, at least not at all times. In some applications, an amplifier with unity gain may be used for matching purposes or to minimize loading effects. Unity gain is a gain of one. That is, the output signal of a unity gain amplifier is at the exact same amplitude as the input signal.

Some amplifiers function as attenuators, with negative gain. The term negative gain means that the gain is less than one (unity). Attenuation gains are usually listed as a fraction, such as 0.1 or 0.025. In this case, the amplitude of the output signal is actually less than the amplitude of the input signal.

Ideally, an amplifier would treat all signals with absolute equality, regardless of the frequency of the input signal. No amplifier circuit has a truly flat frequency response for all possible signal frequencies. Amplifier circuits (including those contained within amplifier ICs) are designed to give the best performance for a specific range of signal frequencies. Typically, amplifier ICs are categorized according to their intended frequency range.

Radio frequency (Rf) amplifiers handle the very high frequencies used in radio receivers and transmitters. Closely related to the rf amplifier is the video amplifier (used in television systems), which operates on similar signal frequencies, but normally has a wider bandwidth to accommodate the wide frequency span of most video signals.

At the opposite extreme is the dc amplifier. A dc amplifier

can boost dc signals. In this case the signal frequency is as low as it can be—0 Hz. Most dc amplifiers can also amplify low-frequency ac signals. The maximum signal frequency will be limited by the specific design of the dc amplifier circuitry.

In general electronics work, the most commonly used type of amplifier is the audio amplifier. This type of amplifier circuit is designed to work on signal frequencies in the audible spectrum, which runs from 20 Hz to 20 kHz (20,000 Hz). Audio amplifiers are used in stereos, intercoms, telephone equipment, television sets, radios, tape recorders, musical instruments, PA systems, and countless other applications.

A great many audio amplifier ICs are widely available today. One of the most popular audio amplifier chips is the LM380. This device is commonly available in two types of DIP housings. The pin diagram for the 8-pin DIP is shown in Fig. 7-5, while the

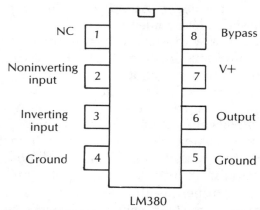

LM380

Fig. 7-5   *The 8-pin LM380 audio amplifier IC is quite popular.*

14-pin version is illustrated in Fig. 7-6. On both of these ICs, only six of the pins are actually used in the circuit. The remaining pins are shorted to ground and provide an internal heat sink. The 14-pin version has six heat sink pins, while the 8-pin version has just two. This means the 14-pin LM380 has more internal thermal protection; therefore, it can handle greater amounts of power without overheating. Of course, an external heat sink can be used with either version of the LM380 to increase its power-handling capability.

Without an external heat sink, the basic LM380 audio ampli-

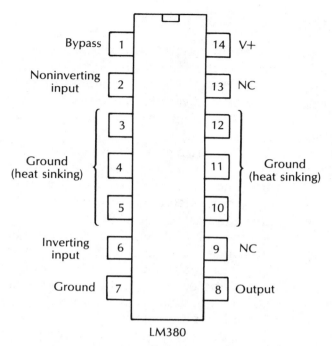

Fig. 7-6 *The 14-pin version of the LM380 audio amplifier IC has extra heat sinking pins.*

fier IC can dissipate up to about 1.25 W at room temperature. If the six heat sink pins of the 14-pin version are soldered to a 6-sq. in. copper foil pad on a PC board (2-oz. foil), the chip can produce up to 3.7 W at room temperature.

The gain of the LM380 audio amplifier IC is internally fixed at 50 (34 dB). The gain can be modified with external circuitry.

The output from this amplifier chip automatically centers itself at one-half the total supply voltage applied to the IC. This eliminates the possibility of any offset problems. If a symmetrical dual-polarity power supply is used, the output signal will be automatically centered around ground potential (0 V) and will contain no dc component.

Some audio amplifier ICs are designed as preamplifiers (or preamps). A preamplifier is similar to a regular amplifier, except that it will accept a lower amplitude input signal and has less internal noise than a regular amplifier circuit. A preamplifier is used to boost very small signals up to levels that are usable by the main power amplifier stage.

Op amps

Probably the most common type of linear IC is a specialized form of amplifier known as the operational amplifier. This is commonly shortened to op amp. Operational amplifiers were originally designed to perform various mathematical operations in analog circuits. In a way, an op amp is a simple, nonprogrammable analog computer.

Op amp circuits require a great many components. An operational amplifier circuit built from discrete components would be bulky and very expensive. For this reason, op amps weren't used very much in general electronics work until they were available in inexpensive and convenient IC form.

A modern op amp can cost less than one dollar up to a few dollars and will fit comfortably on your thumb. They are so inexpensive and easy to use that countless applications have been found for the once exotic op amp, making it perhaps the most common circuit element in modern electronics.

The de facto standard for op amps is the classic 741 chip. At one time, the 741 was the very height of high technology and was considered a very high-performance device. Today the 741 is relegated to noncritical applications and general experimentation. The technology has improved greatly since this chip was designed. Even though it is a low-grade op amp by modern standards, the 741 still works very well and there continues to be a healthy market for this chip, especially since it typically sells for about fifty cents or less.

Most other op amp ICs are designed to be pin-for-pin compatible with the 741. That is, each pin serves the same function on both the 741 and other op amp chips. This way, you can design and experiment with a circuit using a cheap 741. If the IC gets damaged during the course of your experimentation, it's no great loss. Once the circuit has been finalized, the 741 can be replaced with a higher-grade op amp IC. Naturally, an IC socket is almost essential for this sort of planned substitution.

Operational amplifier ICs such as the 741 are offered in a variety of packaging styles. The 8-pin DIP housing, shown in Fig. 7-7, is probably the most commonly used. Sometimes a 14-pin DIP housing will be used, as illustrated in Fig. 7-8. Some op amp ICs are in 8-pin round cans, as shown in Fig. 7-9.

Most op amps require a symmetrical dual-polarity power supply. That is, both a positive and a negative supply voltage

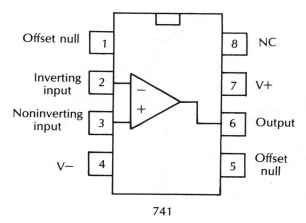

Fig. 7-7   *The 741 op amp is probably the most popular IC. It is shown here in an 8-pin DIP housing.*

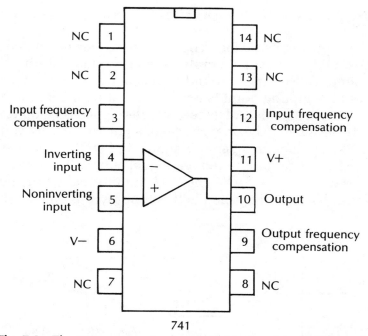

Fig. 7-8   *The 741 op amp is also available in a 14-pin DIP version.*

must be applied to the op amp. These voltages must be symmetrical around ground potential (0 V); the positive voltage should be equal to the negative voltage, except for the reversed polarity, of course.

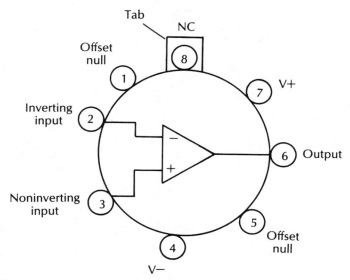

Fig. 7-9 *Some 741 op amps are in 8-pin round can packages.*

The supply voltages for an op amp are sometimes called supply rails. Only the actual supply rails are connected directly to the op amp IC. No direct ground connection is made to the chip itself, but the ground point is normally used with both the input and the output signals. A few recently developed op amp ICs are specially designed to be operated off of a polarity supply voltage.

The symbol for an op amp is shown in Fig. 7-10. Notice that there are two signal inputs and one signal output. The two inputs are marked "+" and "–". The "–" input is called the inverting input. Any signal applied to this input will have its polarity

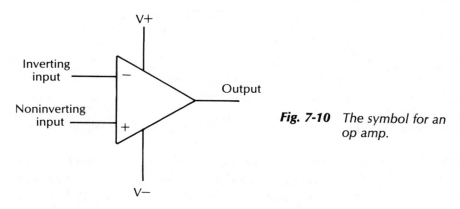

Fig. 7-10 *The symbol for an op amp.*

reversed, or inverted, at the output. If the input signal is positive, the output signal will be negative, and vice versa. An ac input signal will be phase shifted 180 degrees.

The other input to an op amp is marked " + ". This is the noninverting input. The polarity of a signal fed to this input is not reversed, or inverted, at the output. If the input signal is positive, the output signal will also be positive. If the input signal is negative, the output signal will be negative. An ac signal fed to the inverting input of an op amplifier is not phase shifted.

With these two opposing inputs, an op amp is technically known as a differential amplifier. The output is equal to the inverting input voltage minus the noninverting input, multiplied by the amplifier's gain.

In many practical applications, only one of the op amp's two inputs is used. The unused input is normally shorted to ground, so its voltage is effectively zero.

By itself, an op amp has an extremely high gain. Theoretically this open-loop gain is infinite. Of course, no practical component can ever achieve true infinite gain. In practice, the gain is just very, very high. For a 741 op amp the open-loop gain is 200,000. Some higher-grade op amps have even higher open-loop gains.

For most applications, however, this is just too much gain. The output voltage can't exceed the supply voltages, so even a small input signal can saturate the output, resulting in clipping distortion. In most circuits using op amps, the gain is reduced with negative feedback. Some of the output signal is fed back (through a resistance) to the inverting input. This feedback signal is subtracted from the total output signal, so the amplifier's gain is effectively reduced.

Figure 7-11 shows one of the most basic op amp circuits. This is an inverting amplifier circuit. Only the inverting input is used. The noninverting input is shorted to ground, often through a resistor. The polarity of the output signal from this circuit is always the opposite of the polarity of the input signal. If the input signal is a symmetrical ac waveform, it is effectively phase shifted 180 degrees.

Two resistors determine the gain of an inverting amplifier circuit. These are the input resistor (R_i) and the feedback resistor (R_f). The ratio of these two resistances defines the circuit gain,

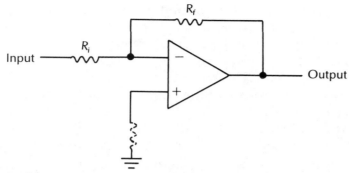

Fig. 7-11   *The inverting amplifier is one of the most basic op amp circuits.*

according to the simple formula:

$$G = \frac{-R_f}{R_i}$$

The minus sign in this equation indicates the polarity inversion of the signal between the input and the output.

If the feedback resistor (R_f) is large in comparison with the input resistor (R_i), the amplifier's gain will be fairly large. On the other hand, if R_f is significantly smaller than R_i, the gain will be very low. In fact, the signal will be attenuated rather than amplified.

If the input resistor and the feedback resistor are equal, the circuit will exhibit unity gain. The output amplitude will equal the input amplitude. Of course, the output signal will be polarity inverted with respect to the input signal. This works even if both resistors are made equal to 0 Ω. That is, the actual resistors can be eliminated from the circuit altogether, as shown in Fig. 7-12. This circuit is known as an inverting voltage follower.

If the noninverting input is used instead of the inverting input, as shown in Fig. 7-13, we have a noninverting amplifier circuit. The output signal will have the same polarity as the input signal. The gain formula for a noninverting amplifier is:

$$G = \frac{1 + R_f}{R_i}$$

Notice that the gain can never drop below 1 (unity) in this type of circuit. Many op amp circuits use both the inverting input and the noninverting input.

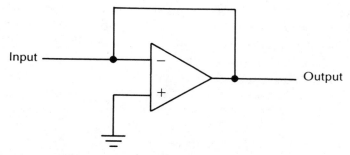

Fig. 7-12 *If no resistors are used in an inverting amplifier circuit, it becomes an inverting voltage follower.*

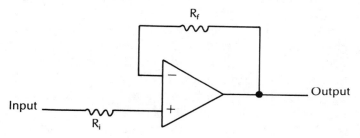

Fig. 7-13 *Another common op amp circuit is the noninverting amplifier.*

Often either the input resistor or the feedback resistor (or perhaps both) will be replaced with some other component, such as a capacitor or a transistor. Of course, this effects the operation of the op amp. Using a capacitor, for example, will give the op amp a definite nonlinear frequency response.

Timers

Another very popular and versatile type of analog IC is the timer. If the 741 op amp (discussed in the preceding section of this chapter) is the most popular chip around today, then the 555 timer IC surely runs a close second.

Timers are used in analog multivibrator circuits. A multivibrator is a circuit with two (and only two) distinct output states, usually called high and low. Obviously, these states are a relatively large voltage (typically close to the circuit's positive supply voltage) and a very low voltage (commonly near ground potential). The multivibrator circuit can switch between these two states almost instantly. In most cases, the actual finite transition time between output states is considered quite negligible and is simply ignored in the majority of practical applications.

Because the output of a multivibrator must always be in one of the two defined states, this type of circuit is a sort of bridge between analog and digital circuits. Many analog multivibrator circuits (including most built around timer ICs) can be used in digital circuitry, even though they are actually linear circuits. Digital circuits will be discussed in chapter 8.

There are three basic types of multivibrator circuits: the monostable multivibrator, the bistable multivibrator, and the astable multivibrator. Timer ICs are commonly employed in monostable and astable multivibrator circuits, but they usually aren't suitable for bistable multivibrator circuits.

A monostable multivibrator has one stable output state. Ordinarily, it stays in this stable state as long as power is applied to the circuit. When a trigger pulse is fed into the monostable multivibrator circuit, the output switches to the opposite, unstable state for a specific period of time, then it reverts to its normal, stable output state until another trigger pulse is received. The length of the output pulse (that is, how long the output stays in the unstable state) is determined by specific resistance and capacitance values in the monostable multivibrator circuit. Typical input and output signals for a monostable multivibrator circuit are illustrated in Fig. 7-14.

A bistable multivibrator has two stable output states. Either output state (high or low) can be held indefinitely, as long as power is applied to the circuit. The only way to get the output to

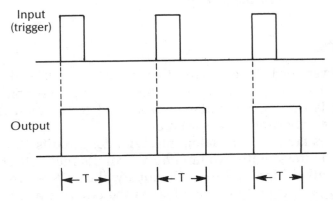

Fig. 7-14 *A monostable multivibrator produces a single fixed-length output pulse each time it is triggered.*

change states is to feed a trigger pulse into the input of the bistable multivibrator circuit. Each time the bistable mutivibrator is triggered, its output jumps to the opposite state. Some typical input and output signals for a bistable multivibrator are illustrated in Fig. 7-15.

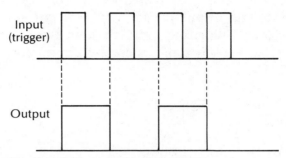

Fig. 7-15 *A bistable multivibrator reverses its output state each time it is triggered.*

Bistable multivibrators are sometimes known as flip-flops because the output flip-flops back and forth when the circuit is triggered. Generally, bistable multivibrators aren't used very often in analog circuitry. They are far more common in digital systems. Most timer ICs cannot be used in bistable multivibrator circuits. This type of multivibrator is described here in the interest of completeness.

Finally, there is the astable multivibrator, which has no stable output states. As long as power is applied to the circuit, the output changes back and forth between its two states at a regular rate. The high time might not be the same as the low time depending on the specific resistance and capacitance values used within the circuit. No input signal or trigger is required for the basic astable multivibrator circuit. A typical output signal from this type of circuit is shown in Fig. 7-16.

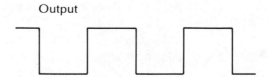

Fig. 7-16 *An astable multivibrator continuously switches between output states.*

Essentially, an astable multivibrator circuit is a rectangular-wave signal generator or oscillator. If the high time and the low time are exactly equal, the output signal is a square wave.

Timer ICs are usually employed in monostable multivibrator and astable multivibrator circuits. By far the most popular timer chip is the 555. The pin diagram of this device is shown in Fig. 7-17. The 555 is usually marketed in an 8-pin DIP housing, as shown here, but occasionally different packagings may be used. If you have a 555 in a package other than an 8-pin DIP, check the manufacturer's spec sheet for the correct pin numbering.

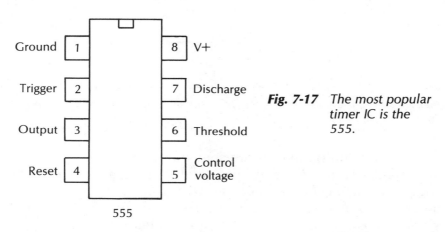

Fig. 7-17 *The most popular timer IC is the 555.*

The basic 555 monostable multivibrator circuit is illustrated in Fig. 7-18. The circuit's timing period (the length of the unstable output pulse) is determined by the values of resistor R_t and capacitor C_t according to the simple formula:

$$T = 1.1R_tC_t$$

The resistance must be in ohms and the capacitance must be farads in order for this equation to work. The 1.1 constant is determined by the internal circuitry of the 555 timer IC.

For reliable operation, the timing resistance (R_t) should be kept within the range of 10 kΩ (10,000 Ω) to about 10 MΩ (10,000,000 Ω). The timing capacitance (C_t) should be kept between 100 pF (0.000000001 F) and 1000 μF (0.001 F).

Figure 7-19 shows the basic 555 astable multivibrator circuit. In this circuit we have one timing capacitor (C_t) and two

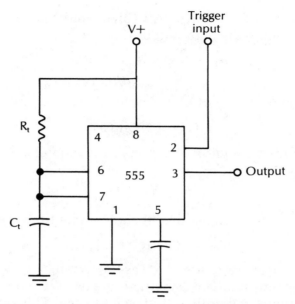

Fig. 7-18   *The basic 555 monostable multivibrator circuit.*

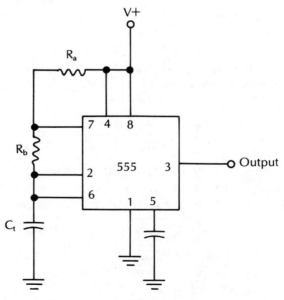

Fig. 7-19   *The basic 555 astable multivibrator circuit.*

timing resistors (R_a and R_b). All three timing components are used to determine the high output time:

$$T_h = 0.693(R_a + R_b)C_t$$

The time the output is in its low state depends only on C_t and R_b; R_a is ignored for this part of the cycle:

$$T_l = 0.693R_bC_t$$

The total output cycle time is simply the sum of the high time and the low time:

$$T_t = T_h + T_l$$

Or, combining these equations, the total cycle time is equal to:

$$T_t = 0.693(R_a + 2R_b)C_t$$

For a recurring ac waveform, such as the rectangular-wave output of an astable multivibrator circuit, we are usually more interested in the frequency, rather than in the cycle time. This isn't a problem because the frequency is equal to the reciprocal of the cycle time:

$$F = \frac{1}{T}$$

Combining these equations, we can find the output signal frequency directly from the component values, using the formula:

$$F = \frac{1.44}{((R_a + 2R_b)C_t)}$$

The duty cycle of the rectangular wave generated by this circuit is the ratio of the high time to the total cycle time. This can be derived directly from the relevant resistor value:

$$T_h:T_t = (R_a + R_b):(R_a + 2R_b)$$

The 556 is a dual timer IC. It has two complete 555 circuits on a single semiconductor chip. The pin diagram for the 556 dual timer is shown in Fig. 7-20.

A quad timer IC, similar to the 555, is also available. This is the 558, shown in Fig. 7-21. The 558 contains four timer sections that are each similar to the 555. These timers are somewhat simplified to limit the number of pins required on the IC. Some 555

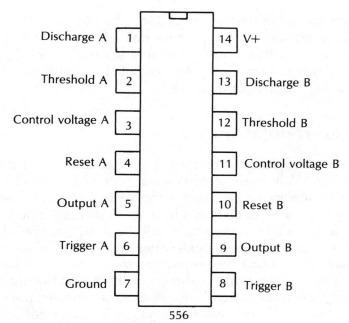

Fig. 7-20 The 556 is a dual 555 timer IC.

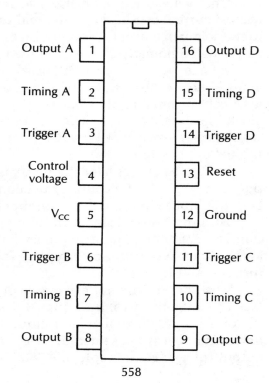

Fig. 7-21 The 558 quad timer IC contains four 555 timer stages.

functions are not available on the 558. The 558 quad timer IC is really intended for use only in monostable multivibrator circuits. It is tricky to use this device in astable multivibrator circuits. A number of other timer ICs are also marketed, but they are all similar in concept to the popular 555.

Voltage regulators

A perhaps mundane, but important, type of linear IC is the voltage regulator. Voltage regulators are very useful in critical circuits that require precise and constant supply voltages. In chapter 5 we saw how the zener diode could be used to regulate a dc voltage, but this approach is rather crude and the results are only fair. A voltage regulator IC does a much better job of rejecting any online transients and holding the output voltage within just a few fractions of a volt of its nominal value. Usually, even if the input voltage drops somewhat below the desired output voltage, the voltage regulator can boost the voltage a little to maintain its nominal output value.

Most voltage regulator ICs also include internal protection against current overload and thermal runaway. If the chip tries to draw too much current, or otherwise gets too hot, the internal circuitry will automatically shut it down.

Voltage regulator ICs aren't nearly as versatile as timers or op amps. There really isn't much in the way of voltage regulator applications outside of power supply circuits. But almost every electronic circuit requires a power supply of some sort, so voltage regulators are very popular and widely used, despite their rather limited applications.

Most voltage regulator ICs have just three leads: input, output, and common. Of course, the common lead is connected to both the input circuit and the output circuit.

With their three leads, voltage regulator ICs look like power transistors, although perhaps somewhat oversized. Similar packaging styles are used. Some typical voltage regulator ICs are illustrated in Fig. 7-22.

Most, but not all, voltage regulator ICs are designed to regulate a specific fixed voltage. The 78xx series is a common example of a fixed-voltage regulator chip. The last two digits in the device number (represented as ''xx'' to generalize) indicate the output voltage. For example, the 7805 is a 5-V regulator, the 7812

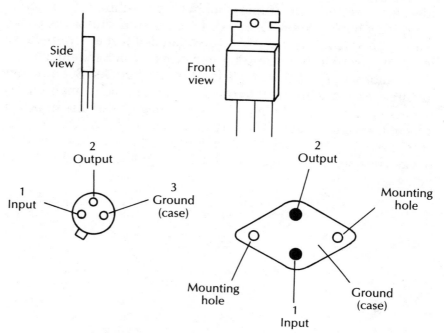

Fig. 7-22 *Some typical voltage regulator ICs.*

has a 12-V output, and the 7815 is a 15-V voltage regulator. The 78xx series is designed for positive voltages (with respect to ground) only. For negative voltages, a similar series is available, numbered 79xx. The 7905 is a – 5-V voltage regulator, and so forth. Some voltage regulators, such as the 723, are variable. The output voltage can be adjusted via external circuitry.

Other linear ICs

There are many, many other linear ICs available, and there is no way we can even begin to cover them all in this short chapter. In fact, it would take more than this entire book just to introduce all the various types of linear ICs. Almost any electronic circuit can be put into IC form, and most have. There are many different types of ICs, and no one can be expected to keep track of them all. Often superficially similar devices have very different pin functions or specifications.

Virtually all manufacturers of ICs publish data books that include full data sheets for the devices they make. It is a good idea for a serious electronics hobbyist or technician to get as many data books from as many different manufacturers as possible. Some are free, while others are offered at just a nominal cost to cover printing and shipping expenses. Other manufacturers charge normal, full book prices for their data books. Such manufacturers are usually not interested in the relatively small orders placed by electronics hobbyists, so I'd suggest that the hobbyist concentrate on the more friendly and reasonable manufacturers. Of course, a professional technician will have to buy the data books that are relevant to whatever equipment and brands are being serviced.

❖ 8
Digital ICs

IN THE LAST COUPLE OF DECADES, MORE AND MORE ELECTRONIC circuits have become digital rather than linear in nature. Digital circuitry makes possible such complex devices as computers.

In a linear (analog) circuit, the electrical signal is usually an analog or direct replica of some external, real-world condition. In a digital circuit, the electrical signal is a numerical representation of some external, real-world condition.

Binary numbers

We are used to counting in the decimal numbering system. This numbering system has ten digits. *Deci-* means ten. Of course, these ten digits are 1, 2, 3, 4, 5, 6, 7, 8, 9, and 0. If we need to express a value larger than nine (the largest available digit), we have to start a new column to the left. Each new column is increased in value by a power of 10. For example

$$41{,}754 = (4 \times 10{,}000) + (1 \times 1000) + (7 \times 100) + (5 \times 10)$$
$$+ (4 \times 1)$$
$$= (4 \times 10^4) + (1 \times 10^3) + (7 \times 10^2) + (5 \times 10^1)$$
$$+ (4 \times 10^0)$$

(Any number raised to the zero power is equal to one.)

This decimal numbering system seems very obvious and natural to us for two reasons. One, we are so used to it; we use it without consciously thinking about it. Second, we happen to have ten fingers.

Ten is the base of the decimal numbering system. This is not the only possible base. Any number can be used as the base. The base defines how many digits the numbering system has. For example, if the base has eight, there would be eight available digits: 1, 2, 3, 4, 5, 6, 7, and 0.

Notice that the digits 8 and 9 are not used in this numbering system. The value eight would be represented as 10_8. The subscript 8 identifies the base of the numbering system. Usually if there is no subscript, the decimal numbering system is assumed, although this might not always be the case. Check the context. Incidentally, a base eight numbering system is called the octal numbering system.

Digital electronics uses the binary numbering system, which has just two digits—1 and 0. Any value greater than one requires an additional column. The decimal and binary numbering systems are compared in Table 8-1. Notice that leading zeros are commonly used in binary numbers. As you can see from this table, a numerical value of any size requires a lot more digits in the binary numbering system than in the decimal numbering system.

Electronic engineers did not decide to use the binary numbering system just to be perverse. While this numbering system is very awkward for human beings to work with, it is very easy for an electronic circuit to use. Each of the two digits can be unambiguously indicated electrically. A low voltage (typically close to ground potential) usually represents a 0, while a 1 is generally indicated by a relatively high voltage (typically just under the circuit's positive supply voltage). In some cases these low and high states may be reversed. No intermediate voltages are allowed in a digital circuit. Either the signal is high or it is low. Because the low state is so close to 0 V, we can say either the voltage is present or not.

Converting between numbering systems is often awkward and tedious. Fortunately, in the vast majority of practical electronics work, you will never need to make such conversions. Usually the circuitry does that for you.

Octal and hexadecimal numbering

In some digital equipment, you might need to enter or read binary values. Unfortunately, it is all too easy to make a mistake in at least one digit when dealing with a value like 110010001110. To make large binary values easier to read, the digits are usually

Table 8 – 1   Comparing the Binary and Decimal Numbering Systems.

Decimal	Binary	Decimal	Binary
0	0000	26	0001 1010
1	0001	27	0001 1011
2	0010	28	0001 1100
3	0011	29	0001 1101
4	0100	30	0001 1110
5	0101	31	0001 1111
6	0110	32	0010 0000
7	0111	33	0010 0001
8	1000	34	0010 0010
9	1001	35	0010 0011
10	1010	36	0010 0100
11	1011	37	0010 0101
12	1100	38	0010 0110
13	1101	39	0010 0111
14	1110	40	0010 1000
15	1111	41	0010 1001
16	0001 0000	42	0010 1010
17	0001 0001	43	0010 1011
18	0001 0010	44	0010 1100
19	0001 0011	45	0010 1101
20	0001 0100	46	0010 1110
21	0001 0101	47	0010 1111
22	0001 0110	48	0011 0000
23	0001 0111	49	0011 0001
24	0001 1000	50	0011 0010
25	0001 1001		

Table 8 – 2   The Octal Numbering System.

Binary	Octal	Decimal	Binary	Octal	Decimal
000 000	0	0	010 010	22	18
000 001	1	1	010 011	23	19
000 010	2	2	010 100	24	20
000 011	3	3	010 101	25	21
000 100	4	4	010 110	26	22
000 101	5	5	010 111	27	23
000 110	6	6	011 000	30	24
000 111	7	7	011 001	31	25
001 000	10	8	011 010	32	26
001 001	11	9	011 011	33	27
001 010	12	10	011 100	34	28
001 011	13	11	011 101	35	29
001 100	14	12	011 110	36	30
001 101	15	13	011 111	37	31
001 110	16	14	100 000	40	32
001 111	17	15	100 001	41	33
010 000	20	16	100 010	42	34
010 001	21	17	100 011	43	35

grouped into sets of three or four. If three digits are used, there
are eight possible combinations, as listed in Table 8-2. In other
words, we are dealing with a variation on the octal numbering
system.

Combining the binary digits into groups of four gives 16 possible combinations, as listed in Table 8-3. This is called the hexadecimal numbering system. Because there are 16 values to be
indicated, and we only have 10 digits, the letters A through F are
used to represent values from 10 to 15.

Table 8 – 3   The hexadecimal numbering system.

Binary	Hexadecimal	Decimal	Binary	Hexadecimal	Decimal
0000 0000	0	0	0001 1010	1A	26
0000 0001	1	1	0001 1011	1B	27
0000 0010	2	2	0001 1100	1C	28
0000 0011	3	3	0001 1101	1D	29
0000 0100	4	4	0001 1110	1E	30
0000 0101	5	5	0001 1111	1F	31
0000 0110	6	6	0010 0000	20	32
0000 0111	7	7	0010 0001	21	33
0000 1000	8	8	0010 0010	22	34
0000 1001	9	9	0010 0011	23	35
0000 1010	A	10	0010 0100	24	36
0000 1011	B	11	0010 0101	25	37
0000 1100	C	12	0010 0110	26	38
0000 1101	D	13	0010 0111	27	39
0000 1110	E	14	0010 1000	28	40
0000 1111	F	15	0010 1001	29	41
0001 0000	10	16	0010 1010	2A	42
0001 0001	11	17	0010 1011	2B	43
0001 0010	12	18	0010 1100	2C	44
0001 0011	13	19	0010 1101	2D	45
0001 0100	14	20	0010 1110	2E	46
0001 0101	15	21	0010 1111	2F	47
0001 0110	16	22	0011 0000	30	48
0001 0111	17	23	0011 0001	31	49
0001 1000	18	24	0011 0010	32	50
0001 1001	19	25			

BCD

It is more and more common for digital circuitry to use BCD
(binary-coded decimal) values. The binary digits are grouped
into sets of four, as in the hexadecimal system, but the upper six

values (10 through 15) are disallowed states and are considered meaningless. Ordinary familiar decimal values can be entered into or read from a BCD system, so the user never has to worry about any other numbering system (see Table 8-4).

Table 8 – 4   Binary – coded decimal.

BCD	Decimal	BCD	Decimal
0000 0000	0	0001 1100	Disallowed state
0000 0001	1	0001 1101	Disallowed state
0000 0010	2	0001 1110	Disallowed state
0000 0011	3	0001 1111	Disallowed state
0000 0100	4	0010 0000	20
0000 0101	5	0010 0001	21
0000 0110	6	0010 0010	22
0000 0111	7	0010 0011	23
0000 1000	8	0010 0100	24
0000 1001	9	0010 0101	25
0000 1010	Disallowed state	0010 0110	26
0000 1011	Disallowed state	0010 0111	27
0000 1100	Disallowed state	0010 1000	28
0000 1101	Disallowed state	0010 1001	29
0000 1110	Disallowed state	0010 1010	Disallowed state
0000 1111	Disallowed state	0010 1011	Disallowed state
0001 0000	10	0010 1100	Disallowed state
0001 0001	11	0010 1101	Disallowed state
0001 0010	12	0010 1110	Disallowed state
0001 0011	13	0010 1111	Disallowed state
0001 0100	14	0011 0000	30
0001 0101	15	0011 0001	31
0001 0110	16	0011 0010	32
0001 0111	17	0011 0011	33
0001 1000	18	0011 0100	34
0001 1001	19	0011 0101	35
0001 1010	Disallowed state		
0001 1011	Disallowed state		

Logic families

Digital ICs are divided into logic families depending on the type of internal circuitry used. The first digital ICs were RTL (resistor-transistor logic) and DTL (diode-transistor logic). Both of these fairly primitive logic families are now almost obsolete. For one thing, they were relatively slow in operation. These types of circuits also permitted only a limited degree of integration. RTL and DTL were suitable for SSI or, at most, MSI.

For a long time the standard digital logic family was TTL (transistor-transistor logic) can operate at quite high speeds and is suitable for reasonably large degrees of integration. That is, very complex TTL digital circuits can be fitted onto a single IC chip, but TTL ICs are very fussy when it comes to their power supply requirements. The supply voltage to a TTL device must be a very well-regulated + 5 V. If the supply voltage drops below about + 4.5 V or goes above + 5.5 V, the TTL chip may not work properly. It could even be damaged or destroyed (even by a too-low supply voltage). But for many years, TTL was the best logic family there was.

TTL devices are commonly identified by a numbering scheme beginning with 74, followed by two or three digits identifying the specific device and its function. For example, a 7400 is a quad NAND gate. (This will be explained shortly.)

As the technology improved, TTL subfamilies were developed. Each new subfamily offered some special advantage. For example, there are high-speed TTL ICs that operate at faster rates than standard TTL devices. The same numbering system was used for the high-speed subfamily with the addition of the letter H in the middle of the component number. For example, 74H00.

Another TTL subfamily is the low-power TTL. The letter L is added to the type number to indicate a low-power device; for example, 74L00.

A 74S00 uses Schottky diodes in its internal circuitry (see chapter 5). This is called a Schottky TTL. Schottky TTL devices can operate at even higher speeds than high-speed TTLs. A low-power Schottky subfamily also exists. It operates at about the same speed as a standard TTL, but consumes much less power. These devices are indicated by the letters LS; for example, 74LS00.

In each case, the middle letter (or letters) indicates the subfamily. The last two (or three) digits indicate the actual device type and function. The 7400, the 74H00, the 74L00, the 74S00, and the 74LS00 all are quad NAND gate ICs. These subfamilies are not always directly interchangeable in a given circuit. Be careful when making substitutions. The various TTL subfamilies are compared in Table 8-5.

While there are still a great many TTL ICs in use, and they are still commonly available, they are gradually being replaced

Table 8 – 5 Comparing the TTL subfamilies.

	Standard TTL	Low – power TTL	High – speed TTL	Schottky TTL	Low – power Schottky TTL
Speed	1	0.1	2	3.5	1
Power consumption	1	0.1	2	2	0.2

by CMOS ICs. The name of this logic family is an acronym for complementary metal-oxide semiconductor. In some technical literature (especially in older materials), this logic family may be referred to as COS/MOS.

CMOS ICs have internal circuitry constructed around *P*-channel and *N*-channel metal-oxide semiconductor transistors (see chapter 6). This logic family offers very low power consumption and the ability to operate from a wide range of supply voltages (typically about + 3 V to + 15 V) at a relatively low cost. Part of this low price, however, is a trade-off in speed. Generally speaking, the higher the power supply voltage (providing it doesn't exceed the chip's maximum limit), the faster CMOS devices can operate. They still tend to be slower than comparable TTL devices. CMOS circuitry is very well suited for LSI devices, as well as SSI and MSI units.

All of the inputs to CMOS gates must always be connected to something. If they are not being used with a digital signal source, they should be tied to either the positive supply voltage or to ground. A floating (unconnected) input can make a CMOS gate quite unstable, causing the output state to become unpredictable. In fact, an unused gate with floating inputs may affect the operation of another gate on the same chip. All input pins on a CMOS IC must be tied either to a suitable digital signal source, to V +, or to ground, whether the individual gate is being used or not.

Many (though not most) CMOS devices are of the tristate type. In addition to the usual logic 0 and logic 1 output states, a tristate circuit might also have a special high-impedance output state, which effectively acts like an open circuit. In other words, a tristate CMOS device can be electrically switched on and off. When it is off (high-impedance output state), it appears to be electrically nonexistent within the circuit.

At least two major numbering systems are used to identify CMOS devices. The 74Cxx system is basically the same as the 74xx numbering scheme used with TTL ICs. A 74C00 performs the same function as a 7400 and the pin layout is the same, but there are significant electrical differences. A CMOS device is never directly interchangeable with a TTL device. In fact, CMOS ICs and TTL ICs normally cannot be used in the same circuit (although there are ways to interface these two logic families, if necessary).

Other CMOS ICs use a completely different numbering system. A four-digit number is used to identify the chip's type and function. This four-digit number always begins with 4, usually (though not always) followed by a zero.

CMOS ICs and static electricity

There is one potential problem to watch out for when working with CMOS ICs. These devices tend to be quite sensitive to static electricity. Stored CMOS chips should have their pins shorted together to prevent damage from an accidental static discharge. Such a static discharge could easily damage the delicate on-chip components. The pins of the CMOS IC can be shorted together for storage by inserting the pins into a special conductive foam (available from many electronics components dealers), or by placing the chips in antistatic plastic containers. Do not use ordinary plastic. If you are unsure if a given container is antistatic plastic, assume it is not. If conductive foam or antistatic plastic containers are unavailable, you can safely store your CMOS ICs by wrapping them in ordinary aluminum foil.

When working with CMOS chips, be very careful and never touch any of the pins unless you are securely grounded. Who hasn't experienced a spark jumping from their fingers to a metal doorknob or other conductive object? Imagine what such a burst of static electric voltage could do to the thin metal-oxide film inside a CMOS chip. Many electronics hobbyists and technicians wear a grounding strap. This is just a snug metal bracelet with a wire attached to it. The free end of the wire is connected to a solid ground point. A cold water pipe is ideal. You can make a home-made grounding strap by clipping a grounding wire to a metal watch band.

If you solder the pins of a CMOS directly, rather than using a socket, your soldering iron should also be properly grounded as insurance against static electric mishaps. Most inexpensive soldering irons are not grounded.

Damage from static electricity was a severe problem with early CMOS devices. Most modern units contain on-chip protection that can help to prevent trouble. I've seen some CMOS chips suffer some pretty strong mishandling and violation of all of these rules, and still work just fine in a circuit. But why take chances? The precautions outlined here are always more than advisable when working with CMOS ICs.

Bits and bytes

Each digital signal (that is, each high or low voltage) is called a bit. The word bit is a compression of binary digit. One bit by itself isn't too useful, so multiple bits are often combined into words. A digital word made up of eight bits is called a byte. If there are only four bits, the word is called a nybble (or nibble).

One-input gates

The basic module of all digital circuits is the gate. A gate is just a simple digital circuit that produces a specific, predictable output condition for each possible input condition. Unlike analog circuits, digital circuits can have just a few, discrete input states. A summary of the output condition for each possible input condition is called a truth table or, in some technical literature, a logic table.

All digital circuits, no matter how complex, are made up of gates. Even the most powerful computer in existence is made up of thousands of individual gates. LSI digital ICs contain multiple gates hardwired into specific configurations. SSI digital ICs are simple gates, with the inputs and outputs individually available to the user. All connections between gates must be made in the external circuitry surrounding the chip.

A digital gate can have any number of inputs and any number of outputs. Obviously, the simplest possible gates have one input and one output. There are two such one-input gates: the buffer and the inverter.

The standard symbol for a buffer is shown in Fig. 8-1. The truth table for this device is given in Table 8-6. The output state of a buffer is always identical to its input state. This might sound like a totally useless gate, and for logic gating purposes, it is rather useless. In practical digital circuits, a buffer is used like a unity gain amplifier in a linear circuit. The buffer provides some isolation between circuit states, preventing later stages from loading down earlier stages. The output of any digital gate can drive just so many inputs to other gates. By using buffers, you can simultaneously drive more gates with a single digital signal.

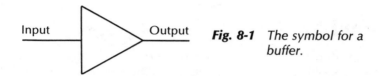

Fig. 8-1   *The symbol for a buffer.*

Table 8 – 6   Truth table for a buffer.

Input	Output
0	0
1	1

An inverter is just the opposite of a buffer. With this type of gate, the output is always in the opposite state as the input. The truth table for an inverter is given in Table 8-7. The standard symbol for an inverter is shown in Fig. 8-2. The small circle at the output indicates inversion (reversal of signal state). Another common name for an inverter is a NOT gate.

Table 8 – 7   Truth table for an inverter.

Input	Output
0	1
1	0

Fig. 8-2   *The symbol for an inverter.*

If we use two inverters in series, as shown in Fig. 8-3, the result will be the same as if we'd used a single buffer. The output state will be the same as the input state. Let's say the input signal is a logic 0. Inverter A will invert this to a logic 1. Then inverter B will invert the signal back to a logic 0. Of course, the same thing will happen if the original input signal is a logic 1.

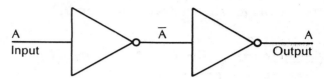

Fig. 8-3 *Connecting two inverters in series results in a buffer.*

Buffers and inverters are usually combined six to a chip. Such ICs are called hex buffers or hex inverters.

By themselves, neither buffers nor inverters are particularly powerful or useful. For most practical digital circuits, more than one input is required.

The AND gate

Figure 8-4 shows the symbol for an AND gate. In its simplest form, an AND gate has two inputs and one output. The single output is controlled by both input signals in a specific way. The output of an AND gate is a logic 1 if, and only if, all inputs are logic 1s. If one or more of the inputs is a logic 0, the output will also be a logic 0. We can see this in the truth table for this device, which is given in Table 8-8. For a two-input AND gate, the output

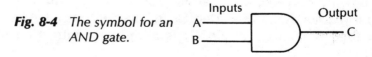

Fig. 8-4 *The symbol for an AND gate.*

Table 8 – 8   Truth table for an AND gate.

Inputs		Output
A	B	
0	0	0
0	1	0
1	0	0
1	1	1

is a logic 1 if, and only if, input A and input B are both 1s. Two input AND gates are usually packaged four to a chip, resulting in a quad AND gate IC.

The principle of the basic AND gate can be expanded to include more than two inputs. It still works in the exact same way. The output is a logic 1 if, and only if, all inputs are logic 1s. The symbol for the three-input AND gate is shown in Fig. 8-5, and the truth table for such a device is given in Table 8-9.

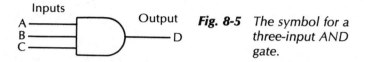

Fig. 8-5 *The symbol for a three-input AND gate.*

Table 8 – 9   Truth table for a three – input AND gate.

Inputs			Output
A	B	C	
0	0	0	0
0	0	1	0
0	1	0	0
0	1	1	0
1	0	0	0
1	0	1	0
1	1	0	0
1	1	1	1

The NAND gate

Suppose we invert the output of an AND gate, as illustrated in Fig. 8-6? In this case, we'd get the exact opposite output we'd get from a regular AND gate. Dedicated inverted output AND gates are called NAND gates. NAND is simply a contraction of not AND.

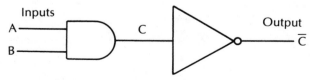

Fig. 8-6   *Inverting the output of an AND gate produces a NAND gate.*

The symbol for a NAND gate appears in Fig. 8-7. The truth table for this type of gate is given in Table 8-10. Notice that the output is a 1 as long as at least one input is a 0. The output is a logic 0 if, and only if, all inputs are logic 1. For a two-input gate, the output is a logic 1 if, and only if, input A and input B are not both 1s.

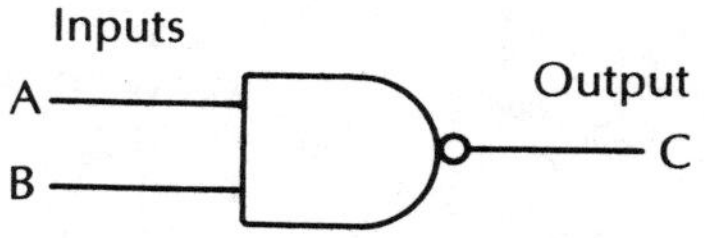

Fig. 8-7 *The symbol for a NAND gate.*

Table 8 – 10   Truth table for a NAND gate.

Inputs		Output
A	B	
0	0	1
0	1	1
1	0	1
1	1	0

The OR gate

Another basic type of digital gate is the OR gate. As you've probably guessed, the name suggests the function. For a two-input OR gate, the output is a logic 1 if, and only if, input A or input B (or both) is a logic 1. The symbol for an OR gate is illustrated in Fig. 8-8. The truth table is given in Table 8-11. Like the AND gate, the basic OR gate can be expanded to accomodate any number of inputs. A three-input OR gate is shown in Fig. 8-9, and the truth table for this device is given in Table 8-12.

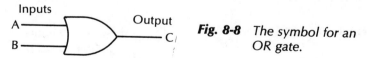

Fig. 8-8 *The symbol for an OR gate.*

Table 8 – 11   Truth table for an OR gate.

Inputs		Output
A	B	
0	0	0
0	1	1
1	0	1
1	1	1

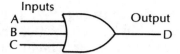

Inputs
A
B
C

Output
D

Fig. 8-9   *The symbol for a three-input OR gate.*

Table 8 – 12   Truth table for a three-input OR gate.

Inputs			Output
A	B	C	
0	0	0	0
0	0	1	1
0	1	0	1
0	1	1	1
1	0	0	1
1	0	1	1
1	1	0	1
1	1	1	1

The NOR gate

If we invert the output of an OR gate, as illustrated in Fig. 8-10, we get a NOR gate, which functions in the exact opposite way as an OR gate. The output of a two-input NOR gate is a logic 1 if, and only if, neither input A nor input B is a logic 1. The symbol for a dedicated two-input NOR gate is shown in Fig. 8-11, and the truth table is given in Table 8-13.

Inputs
A
B

C

Output
$\overline{C}$

Fig. 8-10   *Inverting the output of an OR gate produces a NOR gate.*

Fig. 8-11   *The symbol for a two-input NOR gate.*

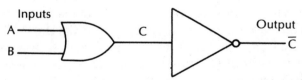

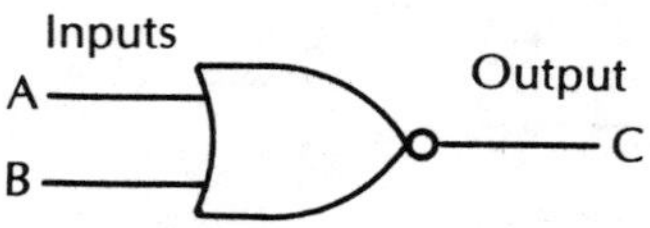

Inputs
A
B

Output
C

Table 8 – 13   Truth table for a NOR gate.

Inputs		Output
A	B	
0	0	1
0	1	0
1	0	0
1	1	0

The X-OR gate

A special type of OR gate is the exclusive-OR gate, or the X-OR gate. The symbol for this device is shown in Fig. 8-12. Unlike the regular OR gate, only one input can be high to produce a high output. The output of the X-OR gate is a logic 1 if, and only if, input A or input B is a logic 1, but not both. The truth table for an X-OR gate is given in Table 8-14. Notice that the output is a logic 1 when the two inputs are at opposite states and a logic 0 when the two inputs are the same. Therefore, the X-OR gate can be considered a digital difference detector or a digital comparator.

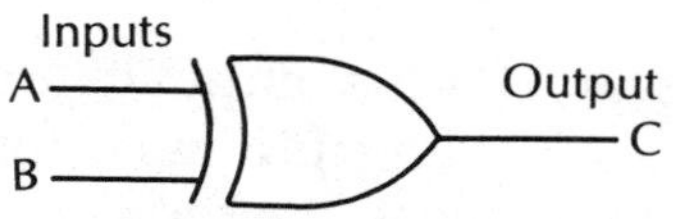

Fig. 8-12 *The symbol for the X-OR gate.*

Table 8 – 14 **Truth table for an X-OR gate.**

Inputs		Output
A	B	
0	0	0
0	1	1
1	0	1
1	1	0

These various types of gates can be combined in an almost infinite variety of ways, producing countless digital circuits and functions.

Boolean algebra

The term Boolean algebra sounds very intimidating and complex, but don't sweat it. If you've read the last few pages of this chapter, you already know 90% of what you'll need to know about Boolean algebra.

Boolean algebra was named after George Boole (1815 – 1864), the mathematician who devised the system. (Boolean algebra is a shorthand way of writing logic combinations, such as we've been using with gates.)

A NOT function is indicated by a bar over the top of the relevant variable, like this:

$$\overline{A}$$

An AND function is indicated by a dot placed between two variables:

$$A \bullet B$$

An OR function is indicated by a plus sign between the two variables:

$$A + B$$

That's all there is to Boolean algebra.

As an example, consider the following Boolean algebra equation:

$$D = A \bullet (B + \overline{C})$$

In this case, D is the output condition, which is high when A and (B or not C). The truth table for this example is given in Table 8-15. The schematic is shown in Fig. 8-13.

Table 8–15 Truth table for the Boolean algebra example.

A	B	C	D
— Inputs			Output

Inputs			Output
A	B	C	D
0	0	0	0
0	0	1	0
0	1	0	0
0	1	1	0
1	0	0	1
1	0	1	0
1	1	0	1
1	1	1	1

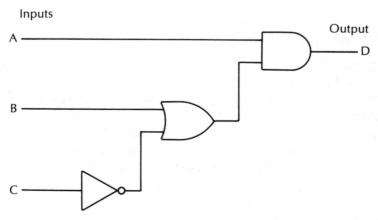

Inputs

Fig. 8-13 *This circuit is used for the Boolean algebra example described in the text.*

Digital multivibrators

Ultimately any digital circuit can be built from individual gates. But for a system of any complexity, this would very quickly become extremly unwieldy. MSI and LSI digital ICs include dozens or hundreds of gates on a single slab of silicon. The user does not have to worry about the interconnections at the gate level, just provide the system inputs and tap off the system outputs.

Many MSI digital ICs are variations on the multivibrator. Multivibrator circuits are somewhat unusual in that they are commonly used in both analog and digital circuits. Although multivibrators were introduced in the last chapter, we will briefly review them here.

A multivibrator is a circuit with two (and only two) distinct output states, usually called high and low. Of course, this is also true of any digital circuit. In a multivibrator, one or both or neither of the output states may be stable. This gives us three basic types of multivibrator circuits:

- The monostable multivibrator,
- The bistable multivibrator, and
- The astable multivibrator.

All three types of multivibrators are used in digital circuits, but the bistable multivibrator is the most commonly used.

A monostable multivibrator has only one stable output state. Ordinarily, it stays in this stable state as long as power is applied to the circuit. When a trigger pulse is fed into the monostable multivibrator circuit, the output switches to the opposite, unstable state for a specific period of time, then it reverts to its normal, stable output state until another trigger pulse is received. The length of the output pulse (that is, how long the output stays in the unstable state) is determined by specific resistance and capacitance values in the monostable multivibrator circuit. The monostable multivibrator is sometimes called a one-shot or a timer. Dedicated digital monostable multivibrator ICs are available, or an analog timer chip, like the 555 (discussed in chapter 7), can be interfaced directly with the digital circuitry.

The astable multivibrator has no stable output states. As long as power is applied to the circuit, the output keeps changing back and forth between its two states at a regular rate. The high time might be the same as the low time depending on the specific resistance and capacitance values used within the circuit. No input signal or trigger is required for the basic astable multivibrator circuit.

Essentially, an astable multivibrator circuit is a rectangular-wave signal generator or oscillator. If the high time and the low time are exactly equal, the output signal is a square wave.

In a digital circuit, an astable multivibrator is often referred to as a clock. The regular pulses are used to synchronize the timing of various stages in moderate to large digital systems.

A bistable multivibrator has two stable output states. Either output state (high or low) can be held indefinitely as long as power is applied to the circuit. The only way to get the output to change states is to feed a trigger pulse into the input of the bistable multivibrator circuit. Each time the bistable multivibrator is triggered, its output jumps to the opposite state. Bistable multivibrators are often called flip-flops because the output flip-flops back and forth when the circuit is triggered.

Generally speaking, bistable multivibrators aren't used very often in analog circuitry. They are far more common in digital systems. Many different digital flip-flop ICs are available.

A simple flip-flop circuit can be made from a pair of NAND gates interconnected as shown in Fig. 8-14. Notice that this circuit has two outputs, labeled Q and $\overline{Q}$ (read Q and not Q). The line over $\overline{Q}$ indicates that this output is inverted. It is the complement (the opposite logic state) of Q. If Q is a 1, $\overline{Q}$ must be 0, and if Q is 0, then $\overline{Q}$ must be 1. This flip-flop circuit also has two logic inputs, labeled S (set) and R (reset). Because of these inputs, this type of bistable multivibrator is known as a set-reset flip-flop or, more commonly, an RS flip-flop.

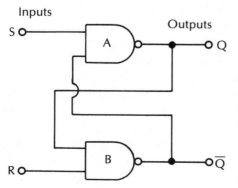

Fig. 8-14 *A simple RS flip-flop can be made from a pair of NAND gates.*

To see how this circuit works, let's first assume that when we first apply power to the flip-flop both S and R are at logic 1 and output Q is 0. Of course, this means that complementary output $\overline{Q}$ must be a logic 1. The inputs to NAND gate A are S (1) and $\overline{Q}$ (1), so the output of this gate (Q) remains at logic 0.

Similarly, NAND gate B's inputs are R (1) and Q (0), so its output $(\overline{Q})$ remains at logic 1. The flip-flop circuit will be latched in this state as long as both inputs (S and R) are kept at logic 1.

If S is now changed to a logic 0, the situation changes. The output of gate A (Q) switches to 1, and this, in turn, forces gate B to change its output $(\overline{Q})$ to 0.

Even if input S is now changed back to 1, the output states will not change because of the way they are cross-fed back to the inputs of the NAND gates. The circuit is latched into this new state, and any changes in the signal on the S input will have no effect on the output signals.

Now, if R is changed to 0, it will change gate B's output $(\overline{Q})$ back to a 1 and gate A's output (Q) back to a 0. Once this is done, further changes in the R input will have no effect on the output. In other words, a 0 input at S sets the flip-flop (Q = 1, $\overline{Q}$ = 0), while a 0 input at R resets the flip-flop (Q = 0, $\overline{Q}$ = 1). A 1 at both inputs will latch the circuit in its present state after the last switch.

Notice that a 0 at both inputs is a disallowed state. The circuit will become confused and the output will be highly unpredictable under this input condition.

The operation of an RS flip-flop is summarized in the truth table of Table 8-16.

Table 8–16   Truth table for an RS flip-flop.

Inputs		Outputs	
R	S	Q	$\overline{Q}$
0	0	Disallowed state	
0	1	1	0
1	0	0	1
1	1	No change — previous state	

Another common type of digital bistable multivibrator is the JK flip-flop. The JK is used simply to distinguish this type of flip-flop from the RS type. These letters don't seem to stand for anything in particular.

The inputs and outputs of a JK flip-flop are illustrated in Fig. 8-15. Again, there are two outputs—Q and $\overline{Q}$. $\overline{Q}$ is always the complement (opposite logic state) of Q. This device also has a total of five inputs—preset, preclear, J, K, and clock.

The preset and preclear inputs on a JK flip-flop work pretty much like the set (S) and reset (R) inputs on an RS flip-flop. A 0 input on the preset terminal immediately forces the Q output to a 1 state (and $\overline{Q}$ to 0). Similarly, a 0 input on the preclear terminal immediately drives the Q output to a 0 state (and $\overline{Q}$ to a 1). Putting both the preset and preclear inputs in a logic 0 state is a disallowed state. If both the preset and preclear inputs are at logic 1, the output signals will be determined by the other three terminals. The operation of the preset and preclear inputs on a JK flip-flop are summarized in the truth table of Table 8-17.

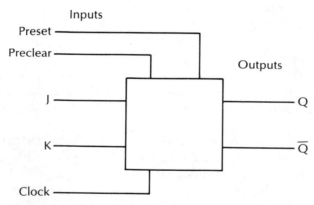

Fig. 8-15 *A more sophisticated type of flip-flop is the JK flip-flop.*

**Table 8 – 17   Truth table for the
preset and preclear inputs of a JK flip – flop.**

Inputs		Outputs	
J	K	Q	$\overline{Q}$
0	0	Disallowed state	
0	1	1	0
1	0	0	1
1	1	Determined by the clocked inputs (see Table 8 – 18)	

The J and K inputs are clocked inputs. This means they can have no effect on the output until the clock input receives the appropriate signal, triggering the flip-flop.

There are two basic types of clocking—level clocking and edge clocking. In a level-clocking system, the flip-flop is triggered by the logic state of the clock input signal. It may be designed to trigger on either a 1 or a 0 (but not both). The flip-flop will remain activated for as long as the clock input is held at the appropriate logic level.

Edge clocking, on the other hand, is triggered by the transition from one clock state to the other. Either the 0 to 1 (positive edge) or the 1 to 0 (negative edge) transition can be used (but not both). Which transition triggers the flip-flop depends on the specific design of its internal circuitry. Obviously, an edge-triggered flip-flop is activated for a much shorter time period than a level-triggered device. For most practical digital work, edge triggering is usually preferred.

Clocked circuits have a number of advantages, especially in moderate to large digital systems. The most important of these is that by triggering all of the subcircuits in the system with the same master clock signal, all operations can be forced to stay in step (synchronization) with each other, preventing many erroneous signals.

The operation of the J and K inputs in a JK flip-flop is summarized in the truth table of Table 8-18. In most practical appli-

Table 8–18   Truth table for the clocked inputs of a JK flip-flop.

Inputs			Outputs	
J	K	Clock	Q	$\overline{Q}$
0	0	N	No change	
0	0	T	No change	
0	1	N	No change	
0	1	T	0	1
1	0	N	No change	
1	0	T	1	0
1	1	N	No change	
1	1	T	Output states reverse (1 becomes 0 and 0 becomes 1)	

N = Clock not triggered
T = Clock triggered

cations, the preset and preclear inputs are not used. They are tied to a permanent logic 1 signal.

The JK flip-flop is quite useful and versatile, and there is no disallowed state for its clocked inputs. But this device's requirement for two inputs in addition to the clock is sometimes inconvenient in certain applications. This problem can be overcome by using a D-type flip-flop. The D stands for data.

Figure 8-16 shows how a D-type flip-flop can be made from a JK flip-flop and an inverter. In this circuit, the J and K inputs will always be in opposite states because of the way they are connected with the inverter. Table 8-19 shows the truth table for a D-type flip-flop. D-type flip-flops also have preset and preclear inputs that function in exactly the same way as in a JK flip-flop. In most practical applications, these inputs are not used.

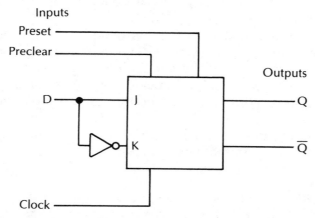

Fig. 8-16   *Another type of flip-flop is the D-type flip-flop.*

Table 8-19   Truth table for a D-type flip-flop.

Input	Outputs	
D	Q	$\overline{Q}$
0	0	1
1	1	0

The D input is functional only when the clock is triggered.

Counters

Since digital circuits deal with the electrical equivalent of numbers, a logical application would be counting. Digital circuit ICs are quite plentiful in today's electronics marketplace.

Digital counters are made up of flip-flop stages. The flip-flops themselves, you should recall, are made up of gates. Therefore, a counter is ultimately derived from basic digital gates.

A supersimple binary counter can be made from a single D-type flip-flop, as illustrated in Fig. 8-17. The $\overline{Q}$ output is fed back into the D input and the count is taken from the Q output.

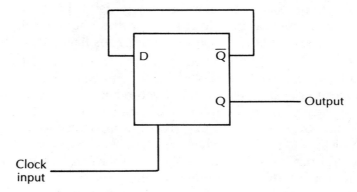

Fig. 8-17   *A one-bit binary counter can be made from a D-type flip-flop.*

In analyzing the operation of this circuit, we'll assume that the Q output starts off as a 1. This means $\overline{Q}$ and D must be at logic 0. Nothing happens until a trigger is received by the clock input, then the flip-flop looks at the data on the D input. Since this is a 0, the truth table tells us that Q should become a 0 and $\overline{Q}$ a 1. This 1 from $\overline{Q}$ is fed back to the D input. By this time, the trigger pulse is gone, so the flip-flop waits until the next trigger pulse comes along. When the clock is triggered, the 1 on the D input changes Q back to 1 and $\overline{Q}$ back to 0, and we're back to where we started.

Typical input and output signals for this circuit are shown in Fig. 8-18. Notice that there is one complete output pulse for every two complete input pulses.

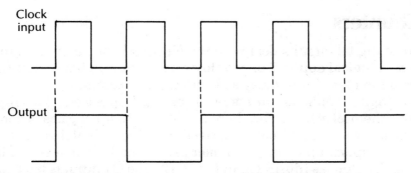

Fig. 8-18 *Typical input and output signals for the circuit of Fig. 8-17.*

This single-stage counter obviously isn't good for much (except for frequency division) because the count resets after a count of 1 is exceeded. The output count runs like this: 0, 1, 0, 1, 0, 1, 0, ...

To create a more useful counter, we can cascade several flip-flop stages to permit a larger count value before resetting. A simple three-stage binary counter is illustrated in Fig. 8-19. Each individual stage functions exactly like the simple single-stage circuit of Fig. 8-17, but each successive stage's clock is the output signal from the preceding stage. In other words, each stage is clocked at one-half the frequency of the preceding stage.

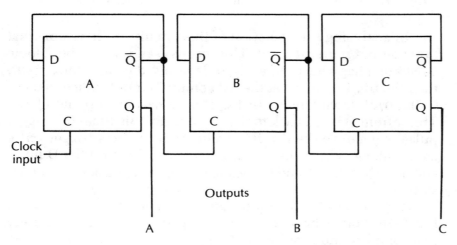

Fig. 8-19 *A simple three-stage binary counter.*

By momentarily grounding all of the preclear inputs, a forced reset can be achieved. The count is restored to 000, regardless of its previous state. The operation of this three-stage counter is summarized in Table 8-20. This basic principle can be easily expanded for even larger counts.

Practical counter ICs usually have more sophisticated circuitry than this simple example. Most offer various special features. Not all digital counter chips count in binary. Many feature built-in circuitry to convert the output count value into hexadecimal, BCD, or even decimal.

Table 8 – 20   Truth table for the three – stage binary counter circuit of Fig. 8 – 19.

Input clock pulse number	Outputs			Decimal equivalent
	C	B	A	
0	0	0	0	0
1	0	0	1	1
2	0	1	0	2
3	0	1	1	3
4	1	0	0	4
5	1	0	1	5
6	1	1	0	6
7	1	1	1	7
8	0	0	0	0 (counter resets)
9	0	0	1	1
10	0	1	0	2
11	0	1	1	3
12	1	0	0	4
13	1	0	1	5
14	1	1	0	6
15	1	1	1	7
16	0	0	0	0 (counter resets)
17	0	0	1	1
18	0	1	0	2
19	0	1	1	3
20	1	0	0	4
21	1	0	1	5

Shift registers

A shift register is a variation on the basic counter circuit. This type of circuit does not necessarily count in a sequential manner, because on any specific clock pulse, the initial D input may be fed either a 0 or a 1 from an external signal source. This input bit is then sequentially passed from stage to stage through the shift register. A shift register can be used as a short-term memory or a digital signal delay.

There are four basic types of shift registers, defined by how the data is fed into and read out of the device. An SISO (serial in, serial out) shift register reads the data in one bit at a time and then feeds it out one bit at a time. A very simple SISO shift register circuit is shown in Fig. 8-20. If there are x stages in the shift register, each input bit will be delayed x clock pulses before it appears at the output.

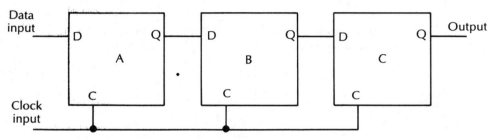

Fig. 8-20   *A simple SISO shift register circuit.*

An SIPO (serial in, parallel out) shift register reads the data in one bit at a time, but has multiple (parallel) outputs from each stage, so that all bits currently in the shift register can be read out simultaneously. A simple SIPO shift register circuit is shown in Fig. 8-21. The operation of this circuit is outlined in Table 8-21.

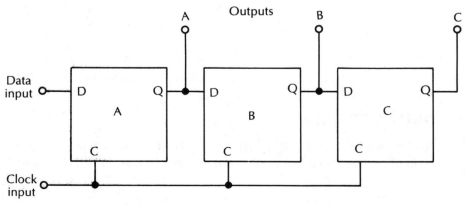

Fig. 8-21   *A simple SIPO shift register circuit.*

Table 8–21   Truth table for an SIPO shift register (all stages initially contain 0s).

	Outputs		
Input	**A**	**B**	**C**
0	0	0	0
1	0	0	0
1	1	0	0
0	1	1	0
1	0	1	1
1	1	0	1
1	1	1	1
0	0	1	1
0	0	0	1
1	1	0	0
0	0	1	0
1	1	0	1
0	0	1	0
0	0	0	1
0	0	0	0

A PISO (parallel in, serial out) shift register loads in the data for all its stages simultaneously. It then feeds the data out of the final stage one bit at a time. A simple PISO shift register circuit is illustrated in Fig. 8-22. This type of shift register is probably the least commonly used.

A PIPO (parallel in, parallel out) shift register loads in the data for all its stages simultaneously, and has multiple (parallel) outputs from each stage so that all bits currently in the shift register can be read out simultaneously. A simple PIPO shift register circuit is shown in Fig. 8-23.

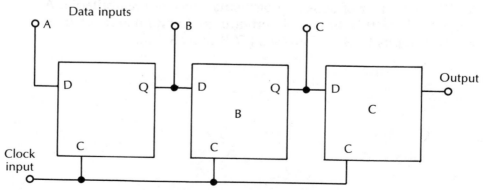

Fig. 8-22   *A simple PISO shift register circuit.*

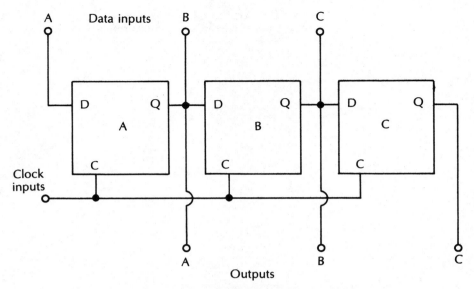

Fig. 8-23   *A simple PIPO shift register circuit.*

Any shift register with a parallel output can be used as if it had a serial output. Just use the single-bit output from the last stage in the register and ignore all of the other output lines.

Other digital devices

There are countless other types of digital ICs. It would be impossible to cover them all here. The digital devices we have discussed in this chapter are the simplest, most versatile, and most commonly used. These are all SSI or LSI devices.

For more information, particularly on sophisticated LSI digital ICs, a number of books on particular devices are available. A good book of this type is *International Encyclopedia of Integrated Circuits* by Stan Gibilisco (TAB Book 3100).

❖Part 3
Miscellaneous components

❖ 9
Transducers

IN THE LAST TWO CHAPTERS OF THIS BOOK, WE WILL LOOK AT some electronic components that don't really fall neatly into either the passive or active classifications. With a few exceptions, these devices are closer to the passive category because they do not amplify.

This chapter is concerned with transducers. A transducer is a device that converts electricity into some other form of energy or some other form of energy into electricity. There are countless different types of transducers, and we can only cover a few of the most important and versatile types in this relatively brief chapter. An entire book could be written just on transducers and it still wouldn't be truly comprehensive.

Crystals

Crystals are transducers that convert mechanical energy (specifically, pressure) into electrical energy, or vice versa. Figure 9-1 shows the basic structure of a crystal as an electronic component. A thin slice of quartz crystal is sandwiched between two metal plates. These plates are held in tight contact with the crystal element by small springs. The entire assembly is enclosed in a metallic case that is hermetically sealed to keep out moisture and dust. Leads connected to the metal plates extend from the case for connection to an external circuit. The symbol for a crystal is shown in Fig. 9-2.

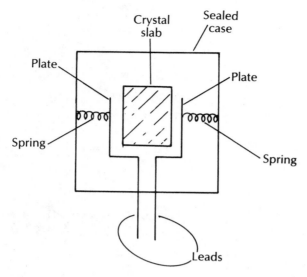

Fig. 9-1 *The basic structure of a crystal.*

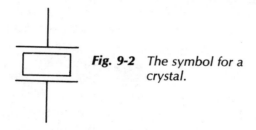

Fig. 9-2 *The symbol for a crystal.*

A crystal works because of a phenomenon called the piezo-electric effect. Two sets of axes pass through the crystal. One set, called the x axis, passes through the corners of the crystal. The other set, called the y axis, is perpendicular to the x axis, but in the same plane. Figure 9-3 shows the two sets of axes in a typical crystal, looking down through the top of the crystal.

Because of the piezoelectric effect, when a mechanical stress is placed across the y axis, a proportional electrical voltage will be produced along the x axis. The reverse also holds true. If an electrical voltage is applied across the x axis, a proportional mechanical stress will be created along the y axis.

A crystal can be used as a pressure sensor, translating me-chanical pressure into a voltage that can be treated by electronic circuitry. Perhaps the most common application for this type of piezoelectric effect is the ceramic cartridge used in many inex-pensive record players.

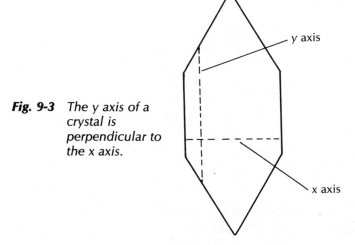

Fig. 9-3 *The y axis of a crystal is perpendicular to the x axis.*

You probably already know that a record is a plastic disc with a spiral groove cut into its surface. This groove undulates in a specific pattern corresponding to the music recorded on the disc. A needle (stylus) is mechanically connected to the crystal element in the ceramic cartridge. This stylus rides in the grooves cut on the record as the disc is spun on a turntable. As the needle is forced back and forth by the physical fluctuations of the groove, different mechanical stresses are put on the crystal. This is converted to an electrical signal that can be amplified and treated by the rest of the circuitry in the record player.

In electronics work, the piezoelectric effect is more often employed in the opposite direction. That is, an electrical voltage is applied across the x axis creating a mechanical stress along the y axis. The piezoelectric effect under these conditions can cause the crystal to vibrate or ring. The vibrations occur (or resonate) at a specific frequency determined by the physical dimensions of the crystal being used.

In practice, only a very thin slice of crystal material is used. This slice may be cut across either an x axis or a y axis. Figure 9-4 shows the equivalent circuit for a typical crystal. Depending on how the crystal is manufactured, it can replace either a series resonant (minimum impedance) or a parallel resonant (maximum impedance) capacitor-coil network. Refer to chapter 4 for more information on resonant circuits. Generally, a crystal designed for series resonant use cannot be used in a parallel resonant circuit, or vice versa.

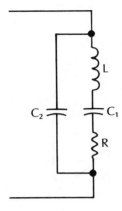

Fig. 9-4 *The equivalent circuit for a typical crystal.*

The advantage of using a crystal in a resonant circuit is that the frequency is very stable and tightly held. The resonant frequency of a capacitor-coil combination may tend to drift under a variety of conditions.

Microphones

A microphone is a transducer that converts audio energy (acoustic sound waves) into electrical energy. There are a number of basic methods for accomplishing this. Most microphone types are represented by the symbols shown in Fig. 9-5. For convenience, the word microphone is often shortened to mike (or mic).

Perhaps the simplest type of microphone is the carbon microphone. Basically, a carbon microphone consists of a small container filled with tiny carbon granules. This container has a carbon disc at either end. One of these discs is rigidly held in a fixed position, while the other disc, called the diaphragm, is movable.

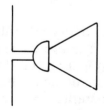

Fig. 9-5 *The symbols for a microphone.*

Sound is caused by fluctuations in air pressure. This varying pressure moves the diaphragm in the carbon microphone back and forth, putting greater or lesser pressure on the carbon particles within the container. The changes in the density of these particles changes their effective resistance. In use this assembly is placed in series with a dc voltage source, such as a battery, as shown in Fig. 9-6. The voltage drop across the microphone will vary along with the changes in the resistance of the carbon particles, which is, in turn, caused by the changing sound pressure striking the diaphragm. In other words, we get a voltage output that varies in step with the sound waves. We have an electrical equivalent of the acoustic energy.

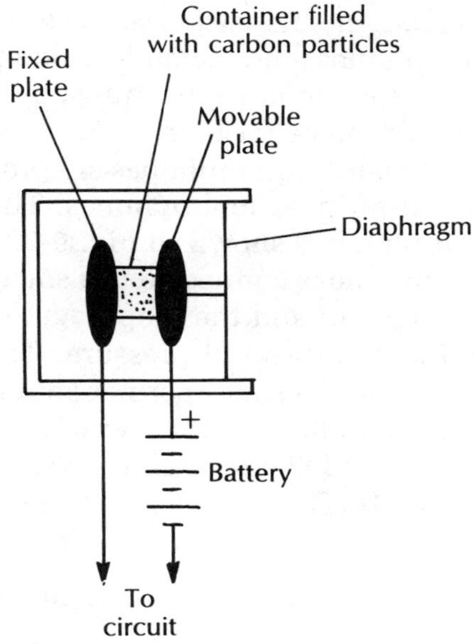

Fig. 9-6 *The structure of a carbon microphone.*

The primary advantages of carbon microphones are that they are relatively low in cost, are relatively durable and sturdy, and they can provide the highest level output of all commonly available microphone types. This means less electronic amplification of the signal is required.

However, there are also a number of significant disadvantages to this type of microphone. Carbon microphones require an external dc voltage source. They have a relatively narrow frequency response, and their noise and distortion levels are higher than for any other common microphone type. Carbon microphones are rarely used in modern electronics.

Another type of microphone employs the piezoelectric effect discussed earlier in this chapter. This is the crystal microphone. The sound pressure on the diaphragm produces a mechanical stress on the crystal element. This generates a proportionate voltage in step with the mechanical stress. While resonant crystals are usually made of quartz, crystal microphones generally use elements made of Rochelle salt.

The crystal microphone offers a number of advantages. It requires no external voltage source, has a fairly high output level, and has a fair frequency response. However, this device is rather fragile. Also the Rochelle salt can absorb moisture, which could ruin it. Both of these problems can be dealt with by replacing the Rochelle salt element with a somewhat more rugged ceramic element. In this case, we have a ceramic microphone. Crystal microphones and ceramic microphones are good for general communications applications, but the frequency response is not really adequate for high-fidelity use.

Dynamic microphones are probably the most popular type of general-purpose microphones. The basic structure of a dynamic microphone is shown in Fig. 9-7. The diaphragm in this type of microphone is connected to a small coil that is suspended so that both the coil and the diaphragm can move freely in response to the incoming sound pressure. The coil moves within the magnetic field of a small permanent magnet. Of course, this induces a voltage in the coil that varies in step with its movement, which is controlled by the sound waves striking the diaphragm (refer to chapter 4). The output voltage level of a dynamic microphone is

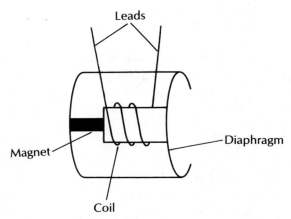

Fig. 9-7 *The structure of a dynamic microphone.*

fairly low, but its frequency response is quite good, and this type of microphone is pretty sturdy and durable.

A variation on the basic dynamic microphone is the ribbon microphone. In this type of microphone a corrugated aluminum ribbon is moved through the magnetic field of a permanent magnet. A small voltage is induced in the ribbon by this movement.

The output voltage from a ribbon microphone is extremely low, and it usually has to be fed through a step-up transformer to reach a usable level. This transformer is often (though not always) contained within the case of the microphone. Even with the transformer, the output level from this type of microphone is very low, but the frequency response is excellent. Ribbon microphones also tend to be quite rugged in construction.

One final type of microphone is the condenser microphone. In this device, two small plates are separated by a small distance. One plate is rigid and the other is flexible (acting as the diaphragm). The movement of the diaphragm varies the distance between the two plates, which changes the capacitance between them (refer to chapter 3). A small circuit within the condenser microphone's case converts this varying capacitance into a proportional varying voltage.

Condenser microphones offer very low distortion and an excellent frequency response. However, they are rather expensive, compared to other types of microphones, and always require their own dc voltage source (usually a small battery).

Speakers

A speaker is the opposite of a microphone. Where a microphone converts sound waves (varying air pressure) into electrical signals, a speaker converts electrical signals into sound-wave patterns of air pressure. A speaker most closely resembles a dynamic microphone. In fact, a small speaker can be used as a low-quality dynamic microphone. The symbol for a speaker is shown in Fig. 9-8. Compare this to the symbol for a microphone shown in Fig. 9-5. Often the term speaker is shortened to SPKR.

The basic construction of a speaker is illustrated in Fig. 9-9. The electrical signal is applied to a coil of wire called the voice coil. Since the voice coil is suspended within the magnetic field of a permanent magnet, it will move back and forth in step with the applied signal, which controls the magnetic field of the coil.

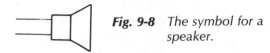

Fig. 9-8 The symbol for a speaker.

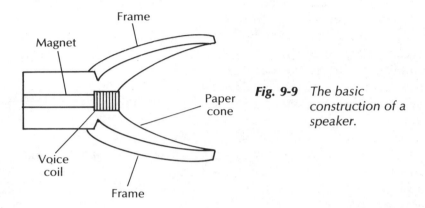

Fig. 9-9 The basic construction of a speaker.

The voice coil is mechanically connected to the center of a paper cone. The outer rim of this cone is firmly attached to a sturdy frame. The paper cone is forced to move in and out along with the motion of the voice coil. This movement of the cone causes corresponding changes in the air pressure. This creates sound waves in step with the original electrical signal applied to the speaker. Some speakers use completely different designs, but this explanation applies to at least 90% of all of the speakers currently in use.

No practical speaker is perfect. All exhibit some signal loss and distortion. Also the frequency response of any practical speaker will always be less than ideal. A small speaker cone will reproduce high frequencies fairly well, because it's low mass permits it to move back and forth rapidly. But a small speaker can't move enough air to reproduce low frequencies well. Conversely, a large speaker cone will do a good job with low frequencies, but it is too massive to handle high-frequency signals well.

This is why most high-fidelity speaker systems contain more than one speaker unit within a single cabinet. A small tweeter reproduces the high frequencies, and a relatively large woofer handles the low frequencies. Sometimes a third speaker will be added to better cover the middle frequencies between the highs of the tweeter and the lows of the woofer. This intermediate speaker is called a midrange speaker. A crossover network or special filtering circuit is used to separate the signal frequencies and feed them to the appropriate speaker unit.

Thermistors

In most electronic components, there is some sensitivity to the ambient temperature surrounding the component. If the component is heated enough, it will change value. In most cases this is undesirable, and high-quality components are often designed to minimize such thermal-sensitive effects.

A thermistor, however, is a component that is specifically designed to respond to changes in temperature. Basically, a thermistor is a resistor whose resistance value changes in a predictable and consistent way with changes in temperature. The symbol for this device is shown in Fig. 9-10.

Fig. 9-10   *The symbol for a thermistor.*

Some obvious applications for thermistors include electronic thermometers and thermostats for precise temperature control. They can also be used in fire alarms. Sometimes thermistors are used to protect other circuitry from damage due to overheating. If the thermistor senses a temperature above a specific, preset level, it triggers a shutdown circuit to remove power from the system until the dangerous temperature condition is eliminated.

LCDs

The LEDs discussed in chapter 5 could also fit in this chapter. They are transducers because they convert electrical energy into light energy. LEDs are used to provide convenient readouts of various circuit conditions.

More and more, multisegment LED displays are being replaced by LCDs (liquid-crystal displays). Most LCDs include one or more seven-segment displays, similar to those used with LEDs. The seven segments are arranged in a figure eight, which permits any digit from 0 to 9 to be displayed. Some LCDs are designed to display alphabetic characters or other special symbols. There is no particular restriction to the displayed image.

Low-grade LCDs can just display specific images, which are determined at the time of manufacture. Others can display almost any image, depending on the applied voltage signal. Such deluxe LCD displays are used in some small, pocket-sized TV receivers. Today, even color LCD screens are possible at reasonable cost. The overall screen size is still quite limited, but this technological problem will probably be overcome soon.

A basic LCD is formed by two pieces of polarized glass with a special fluid sealed between them. The preset images are encoded into the polarization patterns embedded in the glass plates. When a voltage is applied to the LCD's terminals, the liquid near the polarized areas changes color and becomes darker, and the image becomes visible. Some LCDs feature a backlight which allows them to be read in dim lighting conditions.

Photosensors

A popular class of transducers are known as photosensors. A photosensor converts light into an electrical parameter. There are several different types of photosensors. One characteristic of semiconductor crystals is that they are sensitive to light. Most semiconductor components are sealed in lighttight housings to avoid unwanted and uncontrolled effects from ambient lighting conditions. But in a photosensor, we want to expose the semiconductor material to light. All photosensors have some kind of lens or other protected opening for this purpose.

Some semiconductor photosensors are constructed around *PN* junctions. Others are made from a single piece of semiconductor material and are, therefore, junctionless. A junctionless semiconductor photosensor is generally known as a photocell, but this name is potentially confusing because there are two very different types of photocells. They are the photovoltaic cell and the photoresistor.

A photovoltaic cell is a simple slab of semiconductor crystal, usually made of silicon. The semiconductor material is spread out onto a relatively large, thin plate for exposure to as much light as possible. The more of the semiconductor material that is exposed to light, the stronger the electronic response to the light will be.

In a photovoltaic cell, a current flow is created when the cell is exposed to light. In essence, a photovoltaic cell functions something like a dry cell (battery). This is clearly indicated by the symbol for a photovoltaic cell, shown in Fig. 9-11. Often photovoltaic cells are referred to as solar cells or solar batteries, even though the light source doesn't necessarily have to be the sun. Any light source will do.

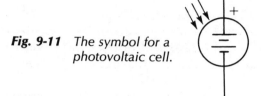

Fig. 9-11   *The symbol for a photovoltaic cell.*

When the silicon surface of a photovoltaic cell is shielded from light, no current will flow through the cell. But when it is exposed to a bright light, a small voltage is generated due to the photoelectric effect. Light striking the semiconductor surface breaks electrons from their atoms and starts them flowing.

If an illuminated photovoltaic cell is hooked up to a load, a current will flow through the circuit. Just how much current will flow depends on the amount of light striking the surface of the photovoltaic cell. The brighter the light, the higher the available current will be.

The photovoltaic cell's output voltage, on the other hand, is relatively independent of the light level. The voltage produced by most photovoltaic cells is about 0.5 V.

Bear in mind that photovoltaic cells, like any other dc voltage source, have a definite polarity. That is, one lead is always positive and the other lead is always negative. These should never be reversed.

The other type of junctionless photocell is the photoresistor, which is sometimes known as the light-dependent resistor. As these names imply, a photoresistor changes its resistance value in proportion to the level of illumination on its surface. Unlike photovoltaic cells, photoresistors generate no voltage themselves. They are very much like the ordinary passive resistors discussed in chapter 2, except their resistance value is light dependent. Photoresistors are usually made of cadmium-sulfide or cadmium-selenide.

Functionally, a photoresistor is a light-controlled potentiom-eter. The light intensity corresponds to the position of the poten-tiometer shaft. Adjusting the light level striking the component adjusts its resistance.

These devices generally cover a pretty broad resistance range—often on the order of 10,000 to 1. The maximum resis-tance—typically about 1 MΩ (1,000,000 Ω)—is achieved when the photoresistor is completely darkened. As the light level increases, the resistance decreases.

A few photoresistors work in the opposite way; that is, the resistance increases with increases in the illumination. But such devices are still quite rare and fairly expensive. At least 95% of all photoresistors feature decreasing resistance with increasing illumination.

Since photoresistors are junctionless devices, like regular resistors, they have no fixed polarity. In other words, a photore-sistor can be hooked up in either direction without affecting cir-cuit operation. Photoresistors can be employed in either dc or ac circuits. Figure 9-12 shows the symbol for a photoresistor.

Fig. 9-12   *The symbol for a photoresistor.*

There are many other types of semiconductor photosensors available. In fact, there are photosensors corresponding to almost every type of semiconductor component, short of integrated cir-cuits. Any standard semiconductor device (with one or more *PN* junctions) can be made photosensitive simply by placing a trans-parent lens in the component's housing so the semiconductor material inside can be exposed to light.

Figure 9-13 shows the symbol for a photodiode. The symbol for a phototransistor is shown in Fig. 9-14. The incoming light acts like a signal on the base lead. Many phototransistors have no physical base lead at all. The light exposure provides the entire base signal. Other phototransistors do have an actual base lead,

Fig. 9-13 *The symbol for a photodiode.*

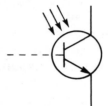

Fig. 9-14 *The symbol for a phototransistor.*

which is usually employed for biasing, while the illumination on the device serves as the actual signal input. Most phototransistors are of the NPN type. While PNP phototransistors do exist, they are rare. In most applications, the exact type of phototransistor is not particularly critical, and substitutions can usually be made freely.

Another type of photosensor is the LASCR (light-activated silicon controlled rectifier). The symbol for this component is illustrated in Fig. 9-15. The incoming light striking the photosensitive surface of the component functions as a gate signal, triggering the LASCR.

Other semiconductor photosensors are also available, but the ones discussed here are, by far, the most commonly used.

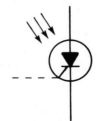

Fig. 9-15 *The symbol for an LASCR.*

Optoisolators

Optoisolators are becoming increasingly popular. With an optoisolator, a signal can be transferred from one circuit to another without any electrical connection between the two circuits. Functionally, an optoisolator is similar to a low-power isolation transformer (see chapter 4), but it is much less expensive and much less bulky. Also, with an optoisolator, you don't have to worry about the effects of the inductive reactance that is inevitable with any transformer.

An optoisolator looks very much like an IC. It consists of a light source (almost always an LED) and a photosensor (any type can be used) enclosed in a lighttight housing. A varying voltage signal is applied to the LED, which causes it to glow with varying brightness levels. The light from the LED shines on the photosensor inducing a corresponding voltage in it. This signal can then be tapped off from the optoisolator's output leads.

Optoisolators can use almost any photosensor as the output device, including (but not limited to) photoresistors, photodiodes, phototransistors, and LASCRs. A typical optoisolator with a phototransistor output is illustrated in Fig. 9-16. The external base lead is optional and may not be included on the optoisolator.

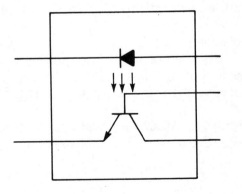

Fig. 9-16 A typical optoisolator with a phototransistor output.

❖ 10
Switches

THIS BOOK BEGAN WITH THE SUBJECT PROBABLY CONSIDERED pretty mundane by many—wires. This concluding chapter is on a type of component many will think is almost as mundane—switches. Yet switches are a vitally important part of the functioning of almost any electronic circuit.

Where a wire conducts electrical signals from one part of a circuit to another, a switch selectively passes or blocks the signal from one part of the circuit to another. This permits the operator to control the functioning of the circuit, even if it's only applying or disconnecting the main operating power. A wire can be considered the roadway for electrical current, while switches serve as traffic exchanges, directing the flow of current.

Switches are generally classified in two basic ways. The first method is to classify the switch by the specific switching functions it performs. The second method of classification is based on the type of construction used in making the switch. We will cover both of these classifications.

Switch functions are defined by the number of poles (circuits controlled) and throws (possible switch positions). There are four basic combinations:

- SPST—single pole, single throw;
- SPDT—single pole, double throw;
- DPST—double pole, single throw; and
- DPDT—double pole, double throw.

A few specialized switches may have more than two poles or throws. This is especially true of rotary switches, which will be discussed later in this chapter. But the vast majority of switches fit into one of these four types.

SPST switches

An SPST switch is the simplest type of switch. It opens or closes a single circuit. Either the controlled current is permitted to flow through the circuit, or the circuit is opened, preventing the flow of current. The symbol for an SPST switch is shown in Fig. 10-1. SPST switches are often used for main power control and for data entry, among other applications.

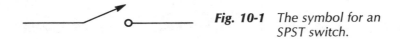

Fig. 10-1   *The symbol for an SPST switch.*

SPDT switches

An SPDT switch is similar to an SPST switch, except its input can be fed to either of two outputs. The symbol for an SPDT switch is shown in Fig. 10-2. Notice that an SPDT switch has three terminals. The middle terminal is called the common connection. This terminal is electrically connected to the movable slider or wiper of the switch. In one position, the switch common is connected to circuit A, while circuit B is left open. In the

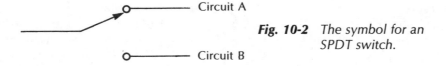

Circuit A

Fig. 10-2   *The symbol for an SPDT switch.*

Circuit B

opposite position, circuit A is open, while the switch common is connected to circuit B.

Some SPDT switches have a center-off feature. If the switch's slider is positioned between A and B, the common will be left unconnected to either of the switched circuits. This type of switch can connect the common to circuit A, or to circuit B, or to neither.

Other SPDT switches do not have center off. With this type of switch, the common must be connected to either circuit A or to circuit B. There are no other possibilities in this case.

An SPDT switch can be used in place of an SPST switch. Simply leave one of the end terminals unconnected, using just the common and the other end terminal. This is illustrated in Fig. 10-3.

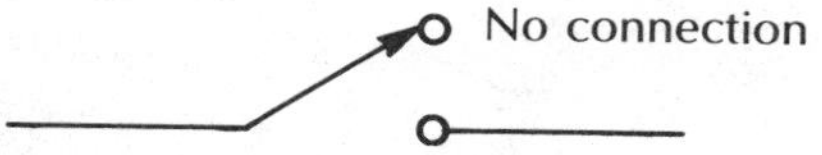

Fig. 10-3 *An SPDT switch can be used in place of an SPST switch.*

No connection

DPST switches

A DPST switch is functionally identical to two SPST switches operated by a single slider. This is illustrated in the symbol for this type of switch, which is shown in Fig. 10-4.

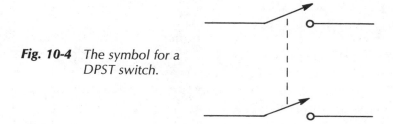

Fig. 10-4 *The symbol for a DPST switch.*

The two poles of a DPST switch can be used in the same circuit or in two entirely separate circuits. There is no internal electrical connection between the two halves (poles) of the switch. The only connection between them is mechanical, forcing the two switches to always operate in unison. Either both switch poles are closed or both switch poles are open, there are no other possibilities.

Dedicated DPST switches aren't too common, although some do exist. In actual practice, if a DPST function is required, a DPDT switch (discussed below) will be used. The terminals for the secondary circuit are simply left unconnected.

DPDT switches

A DPDT switch is like a pair of ganged SPDT switches. This is indicated in the symbol, which is shown in Fig. 10-5. The two halves (poles) of the switch always operate in unison. If common 1 is connected to circuit A, then common 2 is connected to circuit C. If common 1 is connected to circuit B, then common 2 is connected to circuit D.

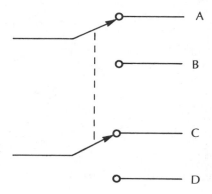

A

B

C

D

Fig. 10-5 *The symbol for a DPDT switch.*

Some (but not all) DPDT switches have a center-off position, as discussed with SPDT switches. In this position, neither common is connected to anything.

The two poles of a DPDT switch can both be used in a single circuit, or they may be used in two entirely different circuits. There is no internal electrical connection between the poles of a DPDT switch. The only connection between the two poles is mechanical, forcing the two poles to operate in unison.

A DPDT switch can be used in place of any of the other basic switch types—SPST, SPDT, or DPST. This is simply a matter of leaving some of the DPDT's terminals unused.

Knife switches

Switches can also be classified according to the type of construction used. The simplest type of switch is the knife switch, shown in Fig. 10-6. Most knife switches are SPST devices, although some SPDT knife switches might be encountered.

An SPST knife switch features a U-shaped metal handle, usually with a plastic knob for the operator to grasp, and two pairs of spring clips. The switch is connected to the external circuitry

A

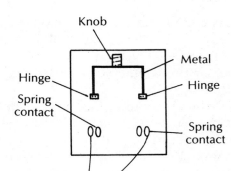

Knob

Metal

Hinge

Hinge

Spring
contact

Spring
contact

To circuit

Fig. 10-6 *A knife switch is a*
very crude and
simple type of
switch: a) switch
open; b) switch
closed.

B

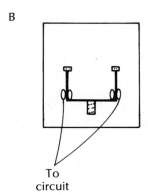

To
circuit

through the spring clips. When the handle is in the position
shown in Fig. 10-6A, the switch is open, and no current can flow
through the attached circuit because there is no complete current
path.

Moving the handle of the knife switch to the position shown
in Fig. 10-6B allows the metal handle to complete the circuit
between the spring clips. The switch is now closed and current
can flow through the circuit.

A knife switch is quite simple and inexpensive to make and
use, but it is rarely used in modern electronics because it is quite
bulky and the electrical connections are fully exposed. This
could mean a serious shock hazard to the circuit operator.

Slide switches

A more commonly used type of switch is the slide switch, illustrated in Fig. 10-7. Slide switches can be made in SPST, SPDT, DPST, or DPDT versions. (In some special designs three or four poles or throws may be available.) Figure 10-7 is an SPST slide switch.

When the slider (the movable portion of the switch) is in the position shown in Fig. 10-7A, the switched circuit is open and no current can flow. But when the slider is moved to the position shown in Fig. 10-7B, a metal strip on the bottom of the slider unit shorts the two terminals together, completing the current path to the external circuit.

Slide switches are inexpensive and relatively easy to use. Their biggest disadvantage is that it is often a big nuisance preparing the rectangular hole required to mount this type of switch on a control panel.

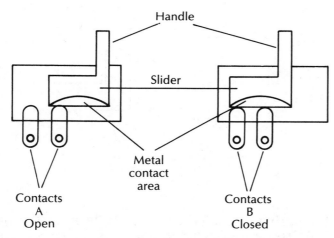

Fig. 10-7 An SPST slide switch.

Toggle switches

Another very popular type of switch is the toggle switch, shown in Fig. 10-8. This type of switch operates in a similar manner to a slide switch, except in this case, the slider unit is in the shape of a ball that rolls in and out of position. Toggle switches are available in SPST, SPDT, DPST, and DPDT versions. It is rare to find a toggle switch with three or more poles or throws.

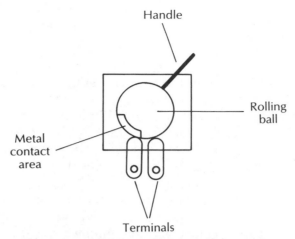

Fig. 10-8   *Toggle switches are quite popular.*

Toggle switches tend to be more expensive than slide switches, but they also tend to be more durable and less susceptible to contamination from dirt and crud in the environment. Also, a toggle switch requires just a simple round hole for mounting.

Push-button switches

The name for our next type of switch is pretty self-explanatory—the push-button switch. A button is pushed to operate this type of switch. Most push-button switches are of the SPST type. Some SPDT push-button switches do exist. DPST and DPDT (or larger) push-button switches are pretty rare.

Some push-button switches are push-on/push-off types. That is, the button is pushed once to close the switch contacts, then pushed a second time to reopen the switch contacts. Either switch position can be held indefinitely.

Most push-button switches, however, are momentary-action types. There are two types of momentary-action switches: normally open and normally closed. Special symbols are used to represent this type of switch. A normally open switch (shown in Fig. 10-9) usually has open contacts. When the button is pushed, the switch is closed for as long as the button is held down. As soon as the button is released, an internal spring-loaded mechanism immediately returns it to its normal open position. The term normally open is commonly abbreviated N.O. or NO.

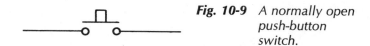

Fig. 10-9   *A normally open push-button switch.*

A normally closed switch (shown in Fig. 10-10) works in just the opposite way as the normally open version. In this case, the switch usually has closed contacts. When the button is pushed, the switch is open for as long as the button is held down. As soon as the button is released, an internal spring-loaded mechanism immediately returns it to its normal closed position. The term normally closed is commonly abbreviated to N.C. or NC.

A few slide switches and toggle switches are spring-loaded for momentary-action operation. Both NO and NC versions are available. Most momentary-action switches, however, are of the push-button type.

Fig. 10-10   *A normally closed push-button switch.*

Rotary switches

When more than two poles or throws are required from a switch, a rotary switch is generally used. This type of switch gets its name from the fact that a knob is rotated to control the position of the slider(s). The symbol used for a rotary switch varies somewhat, depending on the number of poles and throws involved. For example, Fig. 10-11 shows an SP12T rotary switch, with a

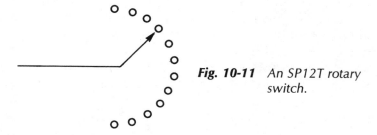

Fig. 10-11   *An SP12T rotary switch.*

single pole and twelve throws. The one slider can be put into any of twelve separate positions.

A 3P6T rotary switch is shown in Fig. 10-12. Here we have three (mechanically interlinked) electrically independent poles,

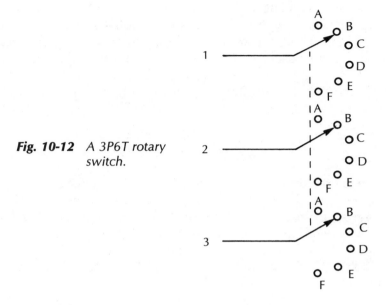

Fig. 10-12 *A 3P6T rotary switch.*

each of which may be placed in any of six separate positions (throws). Note that all of the poles must always operate together; that is, if pole 1 is in position C, then pole 2 and pole 3 must also be in position C. Many other combinations of rotary switches are also possible.

Aside from the pole and throw count, there are two basic varieties of rotary switches. The nonshorting type completely disconnects the circuit at one position before the connection to the next position is made. The other kind of rotary switch is the shorting type, which is also known as the make-before-break type. With this type of switch, when switching from position A to position B, the switch makes contact with both A and B for a brief instant before connection to position A is fully broken and the slider is fully connected to position B. In most circuits, it doesn't really matter which type of rotary switch is used, but some specialized circuits may require one specific type or the other.

Potentiometer switches

Another common type of switch fits onto the back of a potenti-ometer (see chapter 2). Such a potentiometer switch is usually of the SPST type. The switch is mechanically connected to the potentiometer's shaft. When the potentiometer is at its maximum resistance position, the switch is open (off). But as soon as the control knob is advanced a little from this extreme position, the switch's contacts click shut, turning the controlled circuit on. From this point on, the potentiometer functions normally. Potentiometer switches are often used for on-off/volume controls and similar applications.

Relays

A relay is an electrically operated switch. It consists of a coil and a magnetic reed switch in a single housing. When a sufficient voltage is applied across the coil, the generated magnetic field forces the reed switch to open or close, depending on the mechanical design of the individual relay. When no voltage is applied (or if the applied voltage is too low), the switch is in its normal position. Both normally open and normally closed relays are available. Relay switches may be of the SPST, SPDT, DPST, or DPDT type. The symbol for a relay is shown in Fig. 10-13.

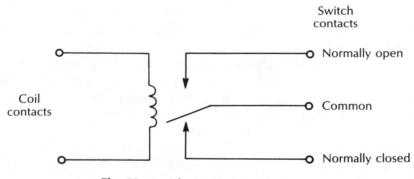

Fig. 10-13 *The symbol for a relay.*

Relays are designed to operate off of specific voltages. Some relays are designed for dc voltages, while others work only with ac voltages.

In some circuits, the resistance across the relay coil may be of significance. This information is usually provided in the manufacturer's specification sheet.

Conclusion

Many other types of electronic components exist, but our space is limited here. Many of the other types of components not covered are really just variations on the basic component types discussed in this book.

We have explored many of the common types of components used in most electronic circuits. The components described here easily account for at least 90% of all components used in all electronics work.

Index

Others Bestsellers of Related Interest

THE THYRISTOR BOOK
—with Forty-Nine Projects—Delton T. Horn

With this collection of 49 projects, Delton T. Horn effectively demystifies the thyristor, explaining in simple terms the theory of thyristor construction and operation. You'll learn the secrets of silicon-controlled rectifiers, triacs, diacs, and quadracs. You'll also get dozens of practical examples for their use, including: light dimmer, self-activating night light, timed switch, visible doorbell, touch switch, and electronic crowbar. 218 pages, 153 illustrations. Book No. 3307, $16.95 paperback, $26.95 hardcover.

THE COMPARATOR BOOK
—with Forty-Nine Projects—Delton T. Horn

Horn sparks new interest in this low-cost device and offers hands-on applications as well as useful in-depth theoretical background. Step-by-step instructions, detailed diagrams, and complete parts lists are provided for each project. You don't have to be an expert, either. The projects gradually become more complex; those presented later in the book build on skills you developed while working on earlier ones. 196 pages, 155 illustrations. Book No. 3312, $15.95 paperback, $23.95 hardcover.

GORDON McCOMB'S GADGETEER'S
GOLDMINE!: 55 Space-Age Projects
—Gordon McComb

This exciting collection of electronic projects features experiments ranging from magnetic levitation and lasers to high-tech surveillance and digital communications. You'll find instructions for building such useful items as a fiberoptic communications link, a Geiger counter, a laser alarm system, and more. All designs have been thoroughly tested. Suggested alternative approaches, parts lists, sources, and components are also provided. 432 pages, 274 illustrations. Book No. 3360, $18.95 paperback, $29.95 hardcover.

49 EASY ELECTRONIC PROJECTS FOR THE
556 DUAL TIMER—Delton T. Horn

Perfect for beginning to intermediate electronics experimenters, this project book features 49 projects designed around the 556 dual timer. The 556 dual-timer IC contains two independent 555-type timers in a single package, making many sophisticated applications possible. Simple, step-by-step building instructions for each applications and many detailed drawings, diagrams, and schematics make tackling these projects easy. 190 pages, 130 illustrations. Book No. 3454, $15.95 paperback, $23.95 hardcover.

ENHANCED SOUND—22 Electronics Projects for
the Audiophile—Richard Kaufman

Get better sound from your audio equipment and save money too! Audio enthusiasts of every level—this is the book you have been looking for! Here are the basics of theory, design, and application you need to improve your audio system without confusing, unnecessary information. Complete with 22 cost-effective electronics projects, this book is designed to boost the performance and capabilities of different types of audio systems. 170 pages, 102 illustrations. Book No. 3017, $11.95 paperback, $15.95 hardcover.

***RADIO-ELECTRONICS®* : From "Drawing Board"**
to Finished Project—Editors of *Radio-Electronics®*

Here's a collection of the "best of the best" projects featured in *Radio-Electronics* magazine! This book has all your favorites—remember the article with plans for building your own sound-effects generator . . . or the one with designs for the sinewave oscillator based circuits . . . or—the list goes on and on, and they've all been included in this book! 160 pages, 80 illustrations. Book No. 3133, $9.95 paperback, $15.95 hardcover.

PRACTICAL TRANSFORMER DESIGN HANDBOOK—2nd Edition—Eric Lowdon

". . . worth its weight in dollar bills."   **—Electron**

From ac-to-dc, dc-to-ac transformation in converter and inverter circuits, and ac-to-dc rectification, the author covers every aspect of transformer design and construction. You'll examine testing transformers and constructing transformers using salvaged parts. 400 pages, 288 pages. Book No. 3212, $34.95 hardcover only.

SUPERCONDUCTIVITY: Experimenting in a New Technology—Dave Prochnow

Now, you have a chance to expand your own knowledge and understanding of superconductivity and to begin performing your working superconductivity experiments. Written for the advanced experimenter, this book includes specific equations and references to the chemistry, thermodynamics, and quantum mechanics that explain this phenomenon. 160 pages, illustrations. Book No. 3132, $14.95 paperback, $22.95 hardcover.

Prices Subject to Change Without Notice.

Look for These and Other TAB Books at Your Local Bookstore

To Order Call Toll Free 1-800-822-8158
(in PA, AK, and Canada call 717-794-2191)

or write to TAB Books, Blue Ridge Summit, PA 17294-0840.

Title	Product No.	Quantity	Price

☐ Check or money order made payable to TAB Books

Charge my ☐ VISA ☐ MasterCard ☐ American Express

Acct. No. _____ Exp. _____

Signature: _____

Name: _____

Address: _____

City: _____

State: _____ Zip: _____

Subtotal $ _____

Postage and Handling
($3.00 in U.S., $5.00 outside U.S.) $ _____

Add applicable state and local
sales tax $ _____

TOTAL $ _____

TAB Books catalog free with purchase; otherwise send $1.00 in check or money order and receive $1.00 credit on your next purchase.

Orders outside U.S. must pay with international money order in U.S. dollars.

TAB Guarantee: If for any reason you are not satisfied with the book(s) you order, simply return it (them) within 15 days and receive a full refund.   **BC**